Numerical Methods for Solving

Differential Equations

D. James Benton

Foreword

If you've struggled to understand the finite element method, then you must read this book. I don't cover a lot of theory in this text. It's mostly a compilation of examples. The one theoretical aspect of numerical methods for solving differential equations that I will present is the finite element method. This powerful technique is more often than not buried under a mountain of esoteric details that make it inaccessible to most students of applied mathematics. I aim to reveal, not obfuscate. I trust you will find this book helpful in that respect.

Understanding applied mathematics results in a richer appreciation of the world and how it works. Those who see only disconnected objects and forces miss the elegance with which such things can be described in the versatile language we call calculus. Differential equations describe the way objects and forces interact. While there are many analytical techniques for solving such problems, this book deals with numerical methods. You must understand some of the former in order to appreciate the latter. Rather than covering the minutia of every obscure method, this book will focus on what works consistently and efficiently. This is a compilation of examples, not a textbook on theory. I trust you will find it interesting and useful.

All of the examples contained in this book,
(as well as a lot of free programs) are available at...
http://www.dudleybenton.altervista.org/software/index.html

Programming

Most of the examples in this book are implemented in the C programming language. Some are contained in Excel® spreadsheets with macros. All of the program source codes—including a triangular grid generator and vectorized matrix solvers—are included. All of the programs will run on any version of Windows® and come pre-compiled. The files are contained in a zipped archive that can be downloaded free of charge at the web address listed above. If you haven't yet stepped up to C, now's the time to do so. C is the most efficient and useful programming language ever conceived.

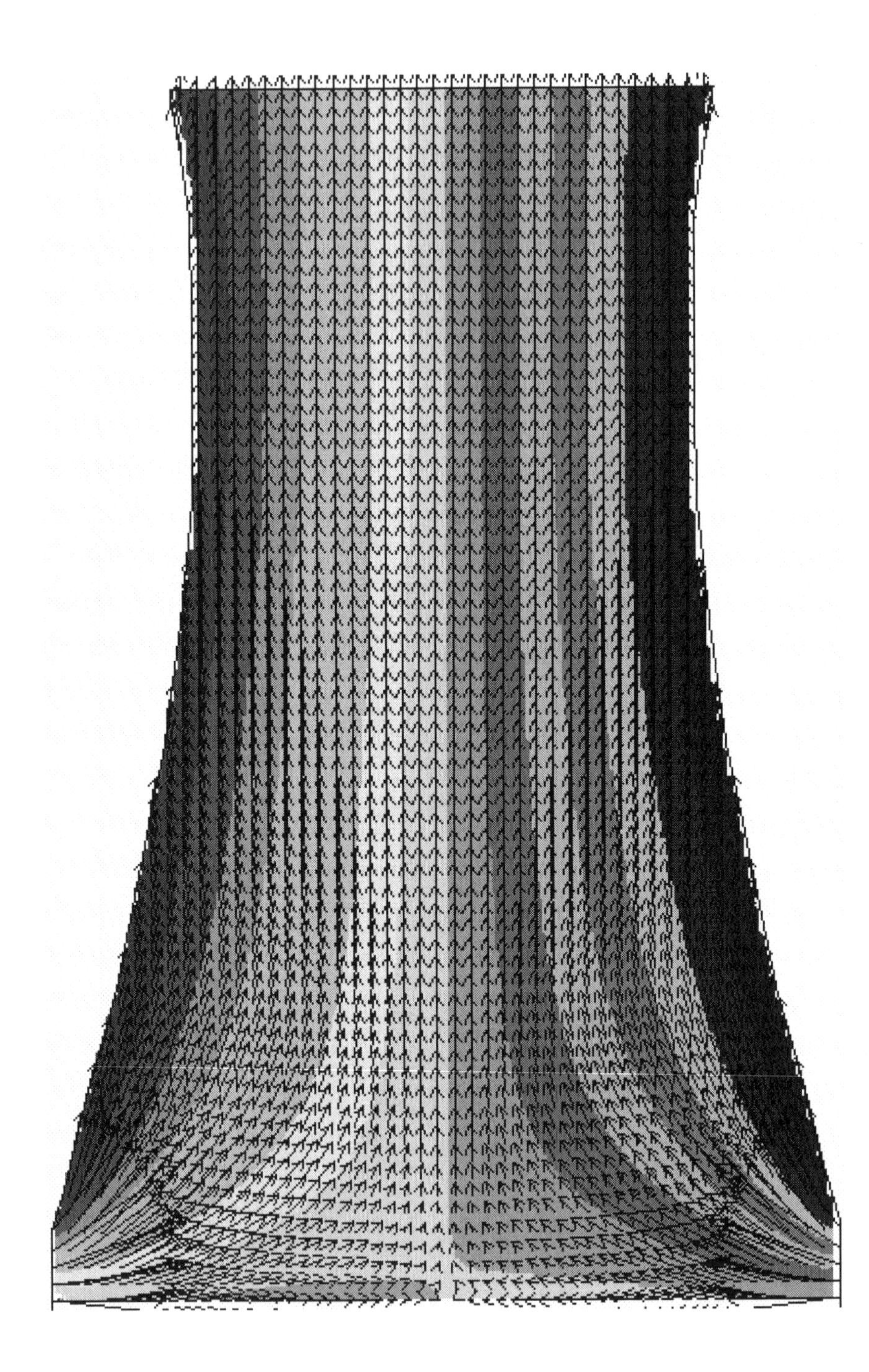

Table of Contents

page

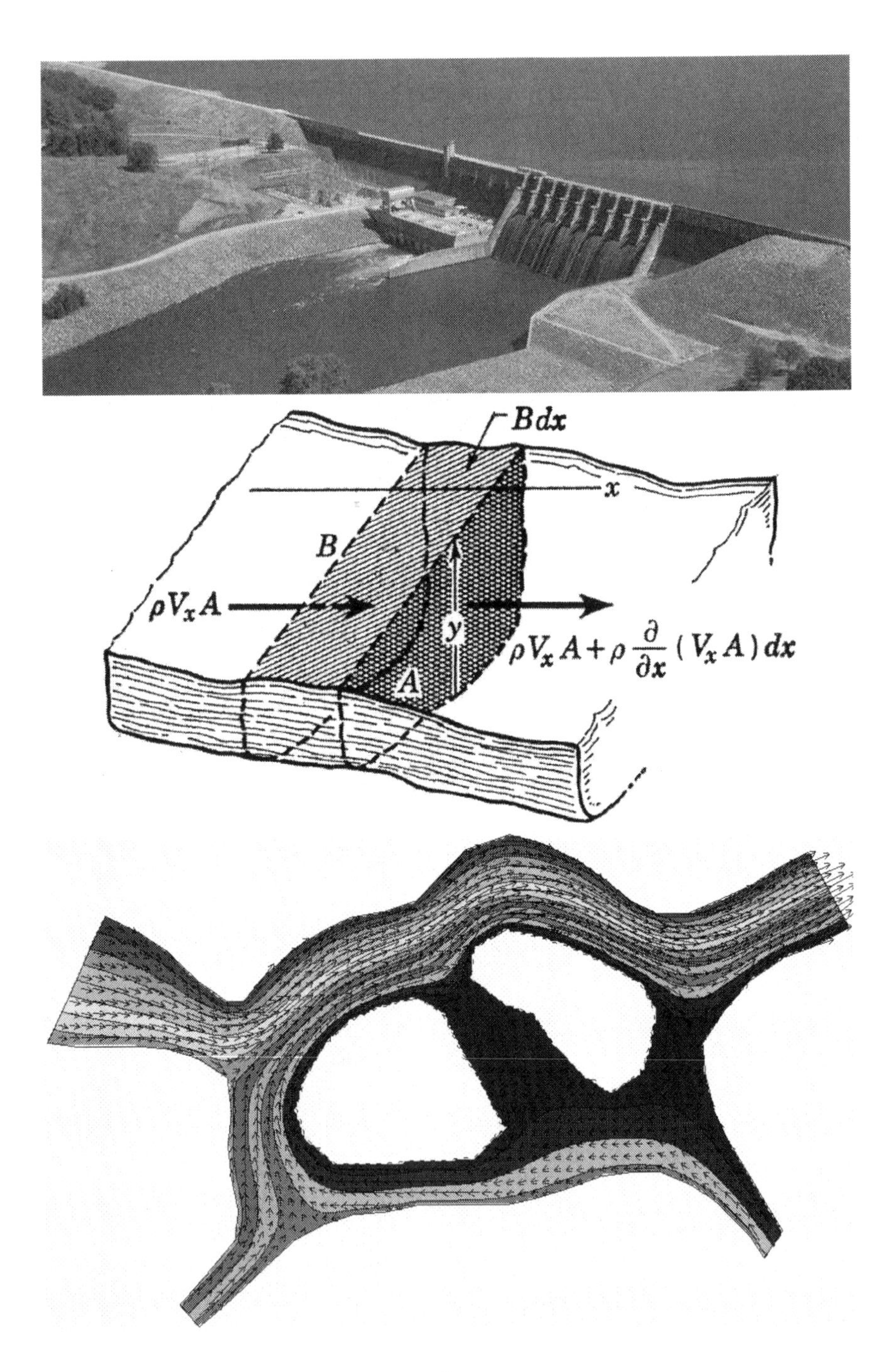

$B\,dx$
x
B
$\rho V_x A$
y
$\rho V_x A + \rho \frac{\partial}{\partial x}(V_x A)\,dx$
A

Chapter 1. Introduction

We will consider four types of differential equations in two groups of two. There are ordinary and partial differential equations. There are also initial value and boundary value problems. The combinations make four groups. There are linear and nonlinear differential equations, but this distinction is of little concern with numerical techniques. There are also individual equations and systems of equations. This distinction is also of little concern here, as we will focus on the methods and provide examples of how to handle these cases. We will consider stiff and loose systems, but these terms could apply to any in the four categories.

Ordinary differential equations basically have one temporal or spatial variable, for example: y(x) or x(t). Partial differential equations have more than one independent variable, for example: y(x,t) or p(x,y). The analytical techniques for solving these are often quite different, so it's not surprising that the numerical techniques would also be different. We begin our discussion with a simple and familiar case: a spring and mass. We could just as easily begin with a pendulum, which we will consider later.

A Simple First Order Ordinary Differential Equation

For the purpose of testing various methods, we will consider a very simple 1st order ordinary differential equation:

$$\frac{dy}{dx} = \sin(x) - \frac{y}{x} \tag{1.1}$$

The analytical solution to this equation is:

$$y = \frac{\sin(x)}{x} - \cos(x) \tag{1.2}$$

Second Order ODE: Damped Harmonic Oscillator

A mass connected to a spring and also a dashpot (damping device) is described by the following differential equation, where *m* is the mass, *c* is the damping factor, *k* is the spring constant, *x* is the displacement, and *t* is time.

$$m\frac{d^2x}{dt^2} + c\frac{dx}{dt} + kx = 0 \tag{1.3}$$

The analytical solution is well known and begins by defining two additional parameters:

$$2\pi f_0 = \omega_0 = \sqrt{\frac{k}{m}} \tag{1.4}$$

$$\beta = \frac{c}{m} \tag{1.5}$$

The parameter $\omega_0=2\pi f_0$ is the resonant frequency (radians/second) and β is the damping coefficient. As many references explain, the solution to Equation 1.3 depends on the value of D in Equation 1.6.

$$D = \beta^2 - 4\omega_0 \tag{1.6}$$

If the determinant D<0 is called under-damped, D>0 is called over-damped, and D=0 is called critically-damped. The under-damped solution is:

$$x = e^{\frac{-\beta t}{2}}\left[A\cos(\gamma t) + B\sin(\gamma t)\right] \tag{1.7}$$

The coefficients A and B are determined by the initial conditions. The parameter γ is given by Equation 1.8:

$$\gamma = \frac{\sqrt{|D|}}{2} \tag{1.8}$$

The over-damped solution is:

$$x = Ae^{\gamma_1 t} + Be^{\gamma_2 t} \tag{1.9}$$

In this case, the paired parameter γ_{12} is given by Equation 1.10:

$$\gamma = \frac{-\beta \pm \sqrt{D}}{2} \tag{1.10}$$

The critically-damped solution is given by:

$$x = (A + Bt)e^{-\omega_0 t} \tag{1.11}$$

Three solutions (ω_0=1, β=0.5,2,3) are shown in the next figure:

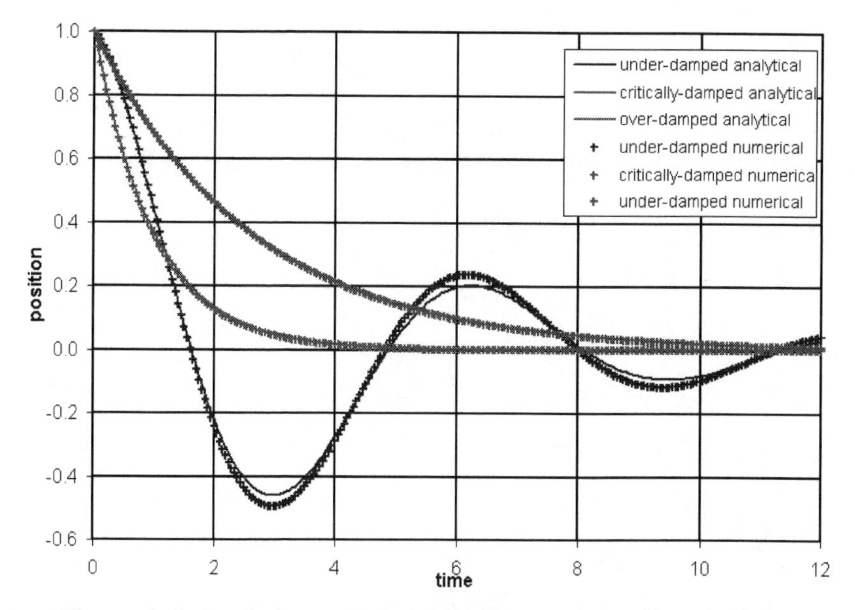

The analytical solutions are the solid lines and the first order (Euler's method) numerical solutions are the points. Agreement is quite good in this case for a time step Δt=0.05 seconds. The curves, calculations, and graph are provided in spreadsheet damped_harmonic_oscillator.xls in the on-line archive. We will use this system and analytical solutions to evaluate various numerical methods in the next chapter.

Third Order ODE

We will now consider the following simple third-order ordinary differential equation:

$$\frac{d^3 y}{dt^3} = -27y \tag{1.12}$$

The initial conditions are:

$$\begin{aligned} y(0) &= 4 \\ y'(0) &= -3 \\ y''(0) &= 9 \end{aligned} \tag{1.13}$$

The solution is:

$$y = 2e^{-3t} + 2e^{\frac{3t}{2}} \cos\left(\frac{3t\sqrt{3}}{2}\right) \tag{1.14}$$

Equation 1.14 plus a simple numerical solution is shown in the following figure:

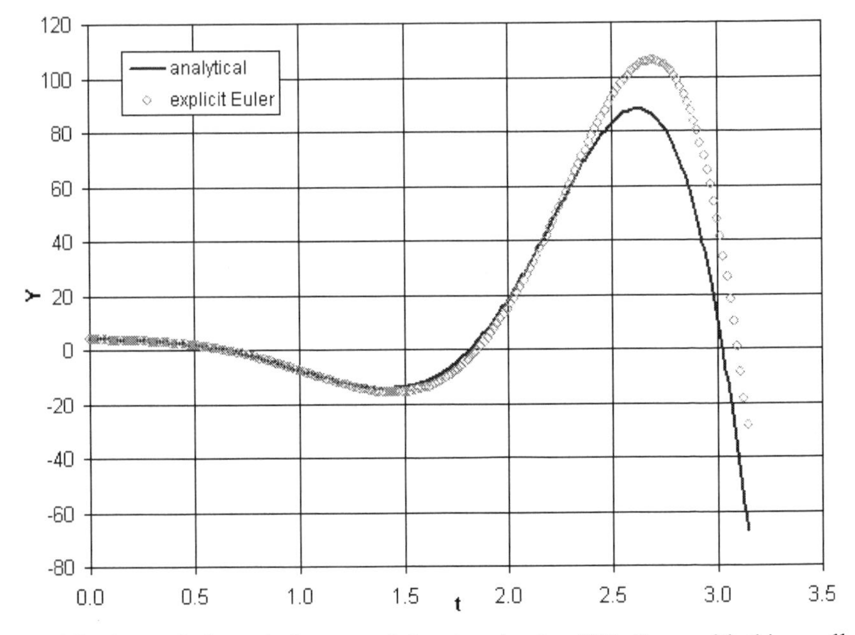

The interval above is 0 to π and the step size is π/200. Even with this small step size, the numerical solution doesn't match well. The higher order methods presented in Chapter 2 will do much better.

Chapter 2. Explicit Runge-Kutta Methods

Runge-Kutta is a type of marching method in that we start with some initial values and then step along through time (or space). We will first consider explicit methods. The seminal reference on the Runge-Kutta and related methods was published by Butcher.[1] There are countless articles on the Web dealing with Runge-Kutta. In marching methods, we consider differential equations of the following form:

$$\frac{dy}{dx} = f(x, y(x)) \tag{2.1}$$

As we shall see, higher order differentials are easily handled by extending this formula. The initial position is represented by x and the time step, Δx, is represented by h. The symbol k is used to represent some particular value of $f(x,y(x))$. The simplest procedure is known as Euler's explicit method, which is implemented:

$$\begin{aligned} k_1 &= f(x, y(x)) \\ y(x+h) &= y(x) + hk_1 \end{aligned} \tag{2.2}$$

This is exactly the same as:

$$y_{x+\Delta x} = y_x + \Delta x \left.\frac{dy}{dx}\right)_x \tag{2.3}$$

Euler's explicit method is sometimes called 1st order Runge-Kutta. In general, these and similar methods can be expressed by the following formula, where n is the number of steps, which is not necessarily the same as the order:

$$\begin{aligned} k_1 &= f(x, y) \\ k_2 &= f(x + c_2 h, y + h(a_{21} k_1)) \\ k_3 &= f(x + c_3 h, y + h(a_{31} k_1 + a_{32} k_2)) \end{aligned} \tag{2.4a}$$

$$\ldots$$

$$k_i = f\left(x + c_i h, y + h \sum_{j=1}^{i-1} a_{ij} k_j\right) \tag{2.4b}$$

$$y = y + h \sum_{i=1}^{n} b_i k$$

[1] Butcher, J. C., The Numerical Analysis of Ordinary Differential Equations: Runge-Kutta and General Linear Methods, John Wiley & Sons Ltd., New York, 1987.

Butcher Tableaus

Butcher expressed the preceding set of equations in tabular form, called a tableau, having the following form:

c_1	a_{11}	a_{12}	a_{13}	...	a_{1n}
c_2	a_{21}	a_{22}	a_{23}	...	a_{2n}
...	...	...	...	...	...
c_n	a_{n1}	a_{n2}	a_{n3}	...	a_{nn}
	b_1	b_2	b_3	...	b_n

(2.5)

The Butcher tableau for Euler's explicit method (Equation 2.2) is:

0	0
	1

(2.6)

We will present all of these methods in this way and then implement them in a code that can handle any formula in this form. There are three common variants of 2nd order Runge-Kutta. The first variant is:

0	0	0
1/2	1/2	0
	0	1

(2.7)

The second variant is called Huen's method:

0	0	0
1	1	0
	1/2	1/2

(2.8)

The third variant is called Ralston's method:

0	0	0
2/3	2/3	0
	1/4	3/4

(2.9)

There are also two common variants of 3rd order Runge-Kutta. The first is:

0	0	0	0
1/2	1/2	0	0
1	-1	2	0
	1/6	2/3	1/6

(2.10)

The second variant of 3rd order Runge-Kutta is:

0	0	0	0
1/3	1/3	0	0
2/3	0	2/3	0
	1/4	0	3/4

(2.11)

There are also two common variants of 4th order Runge-Kutta. The first is:

0	0	0	0	0
1/2	1/2	0	0	0
1/2	0	1/2	0	0
1	0	0	1	0
	1/6	1/3	1/3	1/6

(2.12)

This formula is reminiscent of Simpson's method for numerical integration. The second variant is:

0	0	0	0	0
1/3	1/3	0	0	0
2/3	-1/3	1	0	0
1	1	-1	1	0
	1/8	3/8	3/8	1/8

(2.13)

This formula is reminiscent of Simpson's 3/8ths rule.

Rabiei and Ismail[2] present a 5th order method which has been modified for inclusion here:[3]

0	0	0	0	0	0
0.25	0.25	0	0	0	0
0.25	-0.0086	0.2586	0	0	0
0.5	0.3868	-0.5312	0.6444	0	0
0.75	0.2067	-0.9002	0.8918	0.5517	0
	1.0222	0.0403	-0.107	-0.5999	0.6444

(2.14)

This is the only five-step 5th order method we will consider. Notice that the sum of the absolute value of the coefficients is greater than unity (e.g., 5111>5000). This is a sign that something is wrong. As with Newton-Cotes rules for numerical integration, the low-order methods are stable, but the higher-order methods are not. The stability deteriorates as the order increases.

Rabiei and Ismail also provide the following six-step 5th order method from Butcher:

0	0	0	0	0	0	0
1/4	1/4	0	0	0	0	0
1/4	1/8	1/8	0	0	0	0
1/2	0	-1/2	1	0	0	0
3/4	3/16	0	0	9/160	0	0
1	-3/7	2/7	12/7	-12/7	8/7	0
	7/90	0	32/90	12/90	32/90	7/90

(2.15)

Notice that orders one through four have the same number of steps as the order. When we come to order five, there are six steps (except for the first variant, Equation 2.14). This pattern continues to even higher orders. This is also reminiscent of Newton-Cotes rules (such as Simpson's). All is well with the lower orders, but higher orders seem to fall apart. The same thing happens with

[2] Rabiei, F. and Ismail, F., "Fifth Order Improved Runge-Kutta Methods for Solving Ordinary Differential Equations," *Recent Researchers in Applied Informatics and Remote Sensing*, pp. 129-133, ISBN: 978-1-61804-039-8.

[3] There are several errors in the original publication. The sum of each row must meet the criteria: $c_i=\Sigma a_{ij}$ and $\Sigma b_i=1$, which requires modifying several of the coefficients, including: changing a_{31} from 0.0082 to 0.0086, a_{41} from 0.3860 to 0.3868, a_{42} from +0.5312 to 0.5312, a_{51} from 0.2060 to 0.2067, and b_4 from -0.1 to -0.5999.

Runge-Kutta and for much the same reason. More details on fifth order methods are provided by Luther and Konen.[4] Luther and Konen provide two more variants of 5th order Runge-Kutta. The first is:

0	0	0	0	0	0	0
1/3	1/3	0	0	0	0	0
2/5	4/25	6/25	0	0	0	0
1	1/4	-12/4	15/4	0	0	0
2/3	6/81	90/81	-50/81	8/81	0	0
4/5	6/75	36/75	10/75	8/75	0	0
	23/192	125/192	0	0	-81/192	125/192

(2.16)

The second 5th order variant provided by Luther and Konen is:

0	0	0	0	0	0	0
4/11	4/11	0	0	0	0	0
2/5	9/50	11/50	0	0	0	0
1	0	-11/4	15/4	0	0	0
$\frac{6-\sqrt{6}}{10}$	$\frac{81+9\sqrt{6}}{600}$	0	$\frac{255-55\sqrt{6}}{600}$	$\frac{24-14\sqrt{6}}{600}$	0	0
$\frac{6+\sqrt{6}}{10}$	$\frac{81-9\sqrt{6}}{600}$	0	$\frac{255+55\sqrt{6}}{600}$	$\frac{24+14\sqrt{6}}{600}$	0	0
	4/36	0	0	0	$\frac{16+\sqrt{6}}{36}$	$\frac{16-\sqrt{6}}{36}$

(2.17)

[4] Luther, H. A. and Konen, H. P., "Some Fifth-Order Classical Runge-Kutta Formulas," SIAM Review, Vol. 7, No. 4, 1965.

We will consider four variants of 6th order Runge-Kutta provided by Sarafyan.[5] The first is:

0	0	0	0	0	0	0	0	0
1/9	1/9	0	0	0	0	0	0	0
1/6	1/24	3/24	0	0	0	0	0	0
1/3	1/6	-3/6	4/6	0	0	0	0	0
1/2	1/8	0	0	3/8	0	0	0	0
2/3	-4/3	-21/3	46/3	-29/3	10/3	0	0	0
5/6	-8/72	99/72	-84/72	0	44/72	9/72	0	0
1	107/82	-243/82	0	354/82	-172/82	-36/82	72/82	0
	41/840	0	216/840	27/840	272/840	27/840	216/840	41/840

(2.18)

The second variant of 6th order Runge-Kutta is:

0	0	0	0	0	0	0	0	0
1/9	1/9	0	0	0	0	0	0	0
1/6	1/24	3/24	0	0	0	0	0	0
1/3	1/6	-3/6	4/6	0	0	0	0	0
1/2	1/8	0	0	3/8	0	0	0	0
2/3	6/3	-21/3	16/3	1/3	0	0	0	0
5/6	-68/72	99/72	96/72	-180/72	104/72	9/72	0	0
1	287/82	-243/82	-540/82	894/82	-352/82	-36/82	72/82	0
	41/840	0	216/840	27/840	272/840	27/840	216/840	41/840

(2.19)

[5] Sarafyan, D., "Improved Sixth-Order Runge-Kutta Formulas and Approximate Continuous Solution of Ordinary Differential Equations," *Journal of Mathematical Analysis and Applications*, Vol. 40, pp. 435-445, 1972.

The third variant of 6th order Runge-Kutta is:

0	0	0	0	0	0	0	0	0
1/9	1/9	0	0	0	0	0	0	0
1/6	1/24	3/24	0	0	0	0	0	0
1/3	1/6	-3/6	4/6	0	0	0	0	0
1/2	1/8	0	0	3/8	0	0	0	0
2/3	17/9	-63/9	51/9	0	1/9	0	0	0
5/6	-22/24	33/24	30/24	-58/24	34/24	3/24	0	0
1	281/82	-243/82	-522/82	876/82	-346/82	-36/82	72/82	0
	41/840	0	216/840	27/840	272/840	27/840	216/840	41/840

(2.20)

This last 6th order variant was developed by Huta.[6]

0	0	0	0	0	0	0	0	0
1/9	1/9	0	0	0	0	0	0	0
1/6	1/24	3/24	0	0	0	0	0	0
1/3	1/6	-3/6	4/6	0	0	0	0	0
1/2	-5/8	27/8	-24/8	6/8	0	0	0	0
2/3	221/9	-981/9	867/9	-102/9	1/9	0	0	0
5/6	-183/48	678/48	-472/48	-66/48	80/48	3/48	0	0
1	716/82	-2079/82	1002/82	834/82	-454/82	-9/82	72/82	0
	41/840	0	216/840	27/840	272/840	27/840	216/840	41/840

(2.21)

We will now evaluate these 16 methods (Equations 2.6 through 2.20) against the three analytical solutions presented in Chapter 1. The code to implement these tableaus is listed in Appendix A. The tableaus along with the code to solve the three examples from Chapter 1 can be found in on-line archive in file RKcomparison.c.

[6] Huta, A., "Contribution a la Formule de Sixième Ordre dans la Methodè de Runge-Kutta-Nyström (Contribution to the Sixth-Order Formula in the Runge-Kutta-Nyström Method)," Acta Fat. Rerum Natur. Univ. Cominian. (University Faculty Math Lectures), Vol. 2, pp. 21-23, 1957.

First Order ODE Test

Solving Equation 1.1 from x=0 to x=3 with a step size h=0.1 for each of the 16 methods yields the following table:

equation	order	result	error
1.2	analytical	1.03703	N/A
2.6	1st	1.06468	2.7%
2.7	2nd-1	1.03737	0.0%
2.8	2nd-2	1.03625	-0.1%
2.9	2nd-3	1.03700	0.0%
2.10	3rd-1	1.03703	0.0%
2.11	3rd-2	1.03699	0.0%
2.12	4th-1	1.03703	0.0%
2.13	4th-2	1.03703	0.0%
2.14	5th-1	1.05505	1.7%
2.15	5th-2	1.03703	0.0%
2.16	5th-3	1.03942	0.2%
2.17	5th-4	1.03704	0.0%
2.18	6th-1	1.03703	0.0%
2.19	6th-2	1.03702	0.0%
2.20	6th-3	1.03703	0.0%
2.21	6th-5	1.03703	0.0%

It is not surprising that the 1st order has an error of 2.7%, nor that one of the 2nd order methods (Equation 2.8) has an error of -0.1%. It is surprising that the first 5th order method (Equation 2.14) has an error of 1.7%. This is after fixing several of the terms. Clearly, there is still something wrong with it. We could try other values, but there's no point with so many other methods to choose from. Somewhat surprising is the 0.2% error of the third 5th order method (Equation 2.16). All of the other methods are accurate within 5 to 6 significant figures.

Solving High Order Differential Equations

Differential equations of higher order (i.e., above one) are implemented as an array. The size of the array is equal to the order of the differential equation. The highest order differential goes in the highest index of the array. The next differential is equal to the high order solution. Equation 1.1 becomes:

```
dY[0]=sin(X)-Y[0]/X;
```

Second order ODE Equation 1.3 becomes:

```
dY[1]=-w0*w0*Y[0]-beta*Y[1];
dY[0]=Y[1];
```

Third order ODE Equation 1.12 becomes:

```
dY[2]=-27.*Y[0];
dY[1]=Y[2];
```

```
dY[0]=Y[1];
```

This same pattern is repeated to whatever level is required to specify the differential equation. You can also solve parallel equations having different orders. For example, a third plus a second order differential equation would be entered as:

```
dY[4]=f(x,...);
dY[3]=Y[4];
dY[2]=Y[3];
dY[1]=g(x,...);
dY[0]=Y[1];
```

The first two elements of the arrays (Y[0], Y[1], dY[0], dY[1]) contain the second equation (2nd order ODE) and the next three elements (Y[2], Y[3], Y[4], dY[2], dY[3], dY[4]) contain the second (3rd order ODE).

Second Order ODE Test

The next test is solve Equation 1.3 with the three values of β (0.5,2,3) illustrated in the figure at the end of Chapter 1 using each of the 16 methods. The RungeKutta() function is called 240 times for each value of β. The microsecond timer is called before and after so as to compare the computational effort for each method. Results of this test are shown in the next table:

		h=0.05 (240 steps)			
equation	order	error	μsec	Δμsec	1/Δμsec/err
1.7,9,11	analytical	N/A	90.8	N/A	N/A
2.6	RK1	5.28118000000000	119.6	28.8	0.01
2.7	RK2a	0.08128200000000	146.7	55.9	0.22
2.8	RK2b	0.08128200000000	145.8	55.0	0.22
2.9	RK2c	0.08128200000000	145.8	55.0	0.22
2.10	RK3a	0.00098296500000	185.8	95.0	10.71
2.11	RK3b	0.00098296500000	177.4	86.6	11.75
2.12	RK4a	0.00000958646000	234.7	143.9	725.04
2.13	RK4b	0.00000958646000	234.1	143.3	727.87
2.14	RK5a	3.47677000000000	403.7	312.9	0.0009
2.15	RK5b	0.00000001055510	319.0	228.2	415,091
2.16	RK5c	0.43218100000000	318.2	227.4	0.01
2.17	RK5d	0.00000008270320	316.2	225.4	53,633
2.18	RK6a	0.00000000026026	433.0	342.2	11,227,488
2.19	RK6b	0.01304280000000	434.1	343.3	0.22
2.20	RK6c	0.00000000008336	445.9	355.1	33,784,055
2.21	RK6d	0.00000000007166	433.3	342.5	40,744,544

We subtract the analytical time (90.8 microseconds) from each of the others in order to get the time devoted to the R-K algorithm alone (e.g., 119.6-90.8=28.8 for RK1). We divide 1 by time and error to get a sort of efficiency (shorter time is better and so is less error). All of the 1st, 2nd, and 3rd order methods come out poorly in this test by orders of magnitude. The RK5a, RK5c, and RK6b also come out very poorly, which is why we don't ever want to use any of these methods. They're interesting for theory and history, but useless in practice. A section of the code used to produce this table listed below:

```
double compare1(double b,double*BT,int steps)
  {
  int i;
  double dt=0.05,s,t,x[2];
  beta=b;
  s=t=0.;
  x[0]=1.;
  x[1]=begin();
  for(i=1;i<=240;i++)
    {
    RungeKutta(BT,steps,dxdt,&t,dt,x,2);
    s+=fabs(x[0]-oscillator(t));
    }
  return(s);
  }
void compare3(char*name,double*BT,int steps)
  {
  int j;
  double s;
  __int64 rate,t1,t2;
  QueryPerformanceFrequency(&rate);
  QueryPerformanceCounter(&t1);
  for(s=j=0;j<3;j++)
    s+=compare1(betaj[j],BT,steps);
  QueryPerformanceCounter(&t2);
  printf("%s\t%lG\t%lG\n",name,s,1E6*(t2-
   t1)/((double)rate));
  }
void test2b()
  {
  compare3("analytical",NULL,0);
  compare3("RK1" ,RK1 ,1);
  compare3("RK2a",RK2a,2);
  compare3("RK2b",RK2b,2);
  compare3("RK2c",RK2c,2);
etc...
```

We will now double the time step (h=0.1) and cut the number of steps in half to arrive at the same final result, producing the following table:

		h=0.1 (120 steps)				error
equation	order	error	μsec	Δμsec	1/Δμsec/err	h=0.05/0.10
1.7,9,11	analytical	N/A	97.2	N/A	N/A	N/A
2.6	RK1	5.65166000000000	125.7	28.5	0.01	0.934
2.7	RK2a	0.16463500000000	150.3	53.1	0.11	0.494
2.8	RK2b	0.16463500000000	299.5	202.3	0.03	0.494
2.9	RK2c	0.16463500000000	149.7	52.5	0.12	0.494
2.10	RK3a	0.00398250000000	184.1	86.9	2.89	0.247
2.11	RK3b	0.00398250000000	180.7	83.5	3.01	0.247
2.12	RK4a	0.00007793300000	230.8	133.5	96.09	0.123
2.13	RK4b	0.00007793300000	250.6	153.4	83.66	0.123
2.14	RK5a	3.66533000000000	350.9	253.7	0.0011	0.949
2.15	RK5b	0.00000017534000	570.5	473.2	12,051	0.060
2.16	RK5c	0.43468700000000	334.4	237.2	0.01	0.994
2.17	RK5d	0.00000134235000	360.9	263.7	2,825	0.062
2.18	RK6a	0.00000000853754	454.2	357.0	328,068	0.030
2.19	RK6b	0.02619550000000	525.5	428.3	0.09	0.498
2.20	RK6c	0.00000000275836	460.4	363.2	998,236	0.030
2.21	RK6d	0.00000000237685	442.0	344.7	1,220,423	0.030

The far right column is the error at h=0.05 compared to the error at h=0.1. One might hope that this result would be close to 0.5 (i.e., half the time step yielding half the error). It is for RK2a, RK2b, RK2c, and RK6b, but not so for every method. This brings up an important issue and a new expectation. The best methods are: RK4a, RK4b, RK5b, RK5d, RK6a, RK6c, and RK6d (the same ones that came out on top in the previous test). This is why you hardly ever see anyone implementing any of the other methods. They're only discussed from a theoretical perspective. Using this same table we can further narrow the selection eliminating RK6a in favor of RK6c and RK6d.

Third Order ODE Test

The next test is solve Equation 1.12 over the range 0 to π with a step size of π /50. The results are shown in the following table:

		h=π/50 (50 steps)			
equation	order	error	μsec	Δμsec	1/Δμsec/err
1.14	analytical	N/A	34.9	N/A	N/A
2.6	RK1	1439.7	36.9	2.0	0.00035
2.7	RK2a	52.8459	36.3	2.0	0.0095
2.8	RK2b	52.8459	34.4	2.0	0.0095
2.9	RK2c	52.8459	48.6	13.7	0.0014
2.10	RK3a	2.98706	43.3	8.4	0.040
2.11	RK3b	2.98706	40.5	5.6	0.060
2.12	RK4a	0.140357	61.2	26.3	0.27
2.13	RK4b	0.140357	50.0	15.1	0.47
2.14	RK5a	892.744	99.2	64.3	0.000017
2.15	RK5b	0.000371488	70.7	35.8	75
2.16	RK5c	122.484	69.3	34.4	0.00024
2.17	RK5d	0.00287056	66.8	31.8	11
2.18	RK6a	0.0000394482	98.9	64.0	396
2.19	RK6b	8.62185	89.7	54.8	0.0021
2.20	RK6c	0.0000118713	95.0	60.1	1,402
2.21	RK6d	0.00000997064	90.2	55.3	1,813

A similar pattern as before can be seen here. The same methods prove superior to the rest (RK4a, RK4b, RK5b, RK5d, RK6a, RK6c, and RK6b). In the next chapter we will consider some more interesting applications of these methods.

Chapter 3. Marching Method Applications

In the last chapter we compared the various Runge-Kutta methods and found 7 out of 16 to be superior to the rest (RK4a, RK4b, RK5b, RK5d, RK6a, RK6c, and RK6b). The last two are by far the most efficient. We will now consider some useful examples.

Initial Value Problem: Young-Laplace Capillary Equation

This partial differential equation describes the forces acting on a membrane or the interface between a liquid and vapor. It relates the pressure differential to the curvature:

$$\Delta p = \frac{\sigma}{\frac{1}{R_1} + \frac{1}{R_2}} \tag{3.1}$$

In this equation R_1 and R_2 are the two radii of curvature and σ is the surface tension. The solution we seek is the shape of a sessile droplet.[7] The shape and variables are illustrated in the following figure:

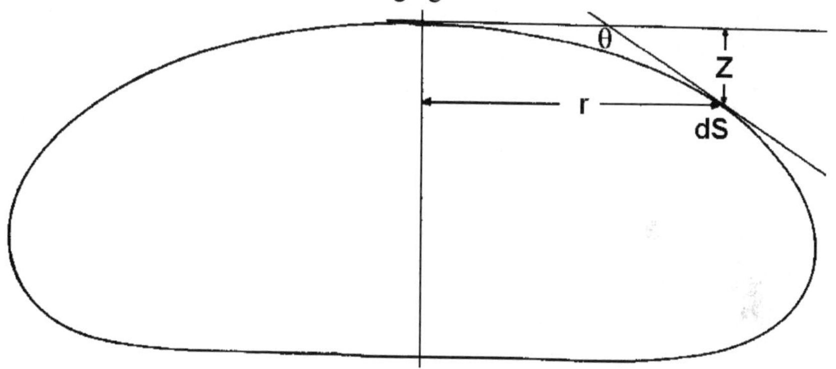

In this figure, *r* is the local radius of the droplet, *z* is the vertical distance, θ us the angle of the surface, and *ds* is the distance along the surface. The initial conditions are r=z= θ=0. Equation 3.1 can be recast in terms of the variables in this figure to become:

$$\frac{dr}{d\theta} = \frac{\cos\theta}{\frac{2}{b} + z - \frac{\sin\theta}{r}} \tag{3.2}$$

---

[7] The web is filled with botanical uses of the word *sessile*, but these have nothing to do with the definition used here. This word comes from the latin *seated*, which is how it's used here.

$$\frac{dz}{d\theta} = \tan\theta \frac{dr}{d\theta} \tag{3.3}$$

$$\frac{dV}{d\theta} = \pi r^2 \frac{dz}{d\theta} \tag{3.4}$$

$$b = \left.\frac{dz^2}{d\theta^2}\right|_{\theta=0} \tag{3.5}$$

The parameter b is a constant equal to the second derivative of the vertical distance at the top/center. Different values of b produce different size droplets. The parameter V is the droplet volume. The vertically projected area is equal to π times the maximum radius squared. We will solve this differential equation using 6th order Runge-Kutta (method RK6d). The equations are entered as follows:

```
#define  R  Y[0]
#define  Z  Y[1]
#define  V  Y[2]
#define dR dY[0]
#define dZ dY[1]
#define dV dY[2]
void drop(double theta,double*Y,double*dY)
  {
  double sr,dS;
  theta=min(theta,M_PI);
  dS=1./b+Z;
  if(fabs(R)>DBL_EPSILON)
    {
    sr=sin(theta)/R;
    if(sr>0.&&sr<1./b)
      dS=2./b+Z-sr;
    }
  dR=cos(theta)/dS;
  dZ=sin(theta)/dS;
  dV=M_PI*R*R*dZ;
  }
```

The source code can be found in the on-line archive in file sessile.c. This has already been compiled and will run on any version of Windows®. There's also a little batch file _compile_sessile.bat to recompile it if desired. Note the 6 #define statements make it easier to keep track of the variables and to express the formulas in the original (non-coded) terms. The results can be found in file sessile.out and sessile.xls. The resulting shapes are shown in this next figure:

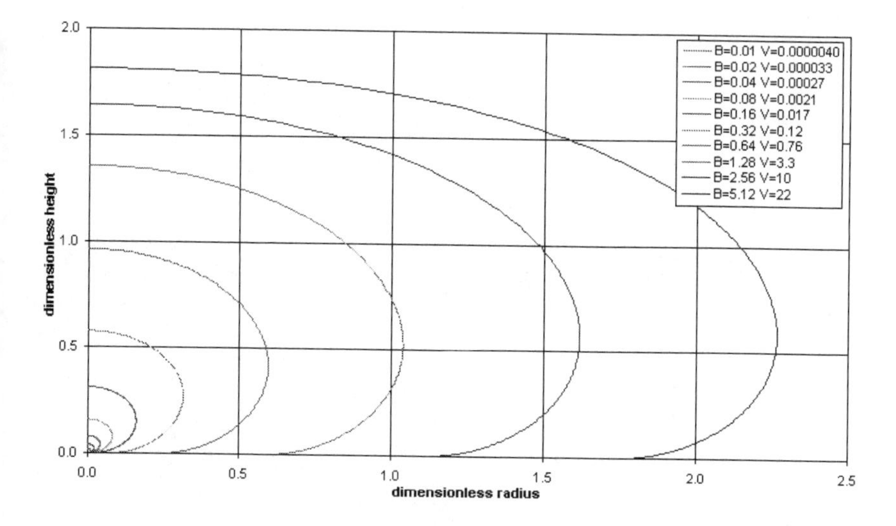

Initial Value Problem: Car Acceleration

No discussion of differential equations would be complete without the first such problem I ever solved numerically: car acceleration. We'll need a torque or horsepower curve:

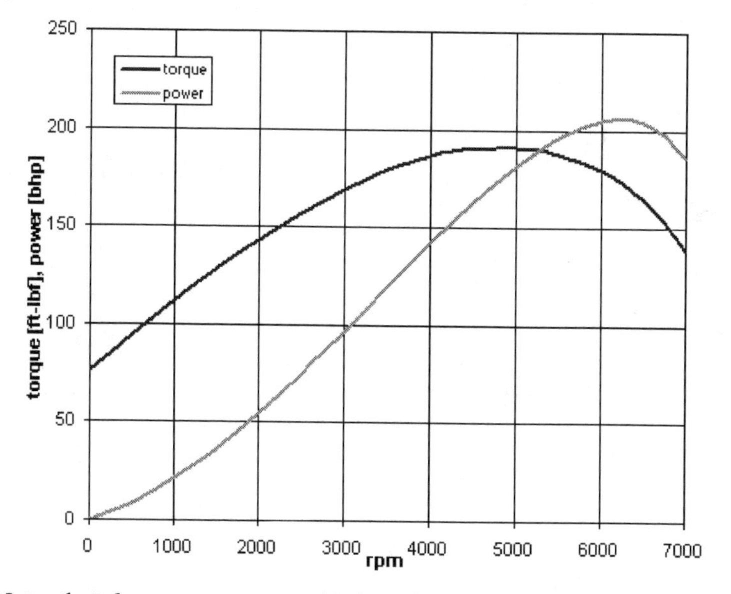

Note that *horsepower=torque*2π*rpm/550/60* so that if you have one curve, the other is easily calculated. We will also need a mass (2861 pounds) and gear ratios:

gear	ratio
1	3.592
2	2.057
3	1.361
4	1.000
5	0.821

We also need to know the axle ratio and tire diameter. The rpm at 60 mph in overdrive will also suffice (2200). We also need to decide the stall and shift speeds (1250 and 6250 rpm). The rest is calculus! Abbreviated results are listed in the following table:

time	rot	torque	gear	accel	veloc	speed	dist
sec	rpm	ft-lbf	shift	ft/sec²	ft/sec	mph	ft
0.00	1250	120.3	1	32.17	0.0	0.0	0
1.00	2305	151.9	1	19.64	21.1	14.4	12
2.00	4796	191.2	1	24.71	43.9	29.9	44
3.00	3968	186.1	2	13.78	63.3	43.2	99
4.00	4846	191.2	2	14.15	77.4	52.8	169
5.00	5722	184.6	2	13.66	91.4	62.3	253
6.00	4278	189.1	3	9.26	103.2	70.4	351
7.00	4664	191.0	3	9.35	112.5	76.7	459
8.00	5052	190.7	3	9.34	121.9	83.1	576
9.00	5437	188.2	3	9.22	131.2	89.4	703
10.00	5814	183.1	3	8.97	140.3	95.7	839
11.00	6178	175.2	3	8.58	149.1	101.6	983
12.00	4760	191.1	4	6.88	156.3	106.6	1136
13.00	4969	191.0	4	6.87	163.2	111.3	1296
13.15	5001	190.9	4	6.87	164.2	112.0	1320

All the details can be found in the on-line archive in spreadsheet car_simulation.xls. This problem is solved in the spreadsheet using the explicit Euler method with a small time step (0.01 sec). You can change values in the spreadsheet and the simulation will update automatically. This problem can also be solved using Runge-Kutta methods. The source code can be found in car_simulation.c. Due to the gear shifting, the implementation is a little more complicated than the previous problems, but still rather simple:

```
double torque(double rpm)
  {
  return ((((((-2.39436962861068E-
    24*rpm+4.94254828375489E-20)*rpm
    -3.88505184269642E-16)*rpm+1.39955406645336E-12)*rpm
    -2.54306117072706E-9)*rpm-1.88969627673033E-19)*rpm
```

```
  +0.037365290797588)*rpm+76.1035566976854;
  }

double horsepower(double rpm)
  {
  return torque(rpm)*rpm*2.*M_PI/550./60.;
  }

double g=32.174;
double weight=2851.;
double rpmat60=2200.;
double rpm_min=1250.;
double rpm_max=6250.;
double ratios[]={3.592,2.057,1.361,1.,0.821};
#define ngear (sizeof(ratios)/sizeof(ratios[0]))

double rpms(int gear,double fps)
  {
  double rpm;
  rpm=rpmat60*(fps/88.)*ratios[gear]/ratios[ngear-1];
  if(rpm>=rpm_min)
    return rpm;
  return rpm_min;
  }

int gears(double fps)
  {
  int gear=0;
  while(fps>0.&&gear<ngear-1)
    {
    if(rpms(gear,fps)<rpm_max)
      break;
    gear++;
    }
  return(gear);
  }

int gear;
double rpm,torq;

void car(double t,double*y,double*dy)
  {
  gear=gears(y[1]);
  rpm=rpms(gear,y[1]);
  torq=torque(rpm);
  dy[1]=min(g,torq*rpm*2.*M_PI/60./max(y[1],
   DBL_EPSILON)/(weight/g));
  dy[0]=y[1];
  }
```

Note that you must handle the case of $V=0$ so as to not divide by zero. The results are shown in the following figure.

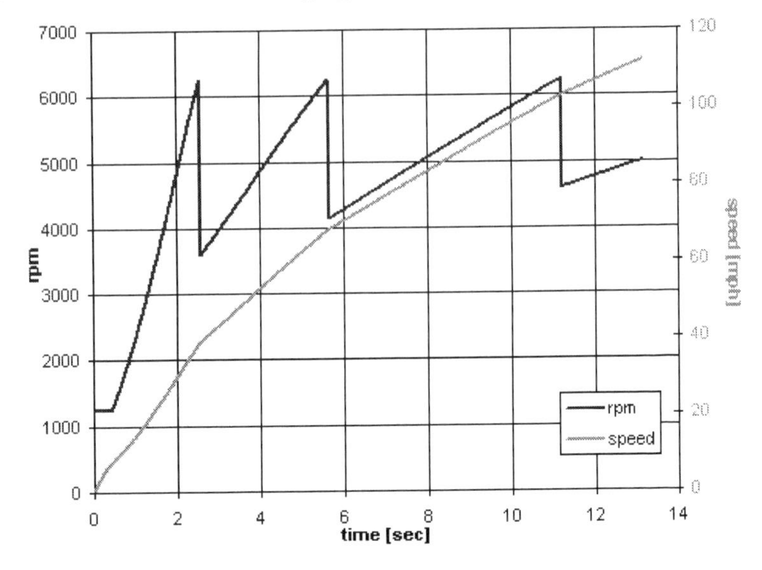

Boundary Value Problem: Blasius Equation

This problem arises from the partial differential equation describing fluid flow near a fixed surface. This near-wall region is call the boundary layer. A detailed discussion of this topic can be found in any fluid mechanics text. A recent reference readily available on-line is the paper by Jaman, Molla, and Sultana.[8] A transformation of variables is applied to the governing equation that results in the following problem:

$$\begin{gathered} 2y''+yy'=0 \\ y(0)=y'(0)=0 \\ y'(\infty)=1 \end{gathered} \tag{3.6}$$

As usual, y' indicated dy/dx and y'' indicates d^2y/dx^2. This is not an initial value problem, rather it is a boundary value problem. Still, it can be solved as an initial value problem by guessing the value of $y'(0)$ that results in $y'(\infty)=1$. Guessing an initial value to arrive at an eventual result is called *shooting*. There

[8] Jaman, M. K., Molla, M. R., and Sultana, S., Numerical Approximations of Blasius Boundary Layer Equation, Dhaka University Journal of Science, Vol. 59, No. 1, pp. 87-90, 2011.

are may problems that can be solved in this way. For more on this the reader is directed to the paper by Summiya.[9]

Equation 3.6 is easily implemented as follows:

```
dY[2]=-Y[0]*Y[2]/2.;
dY[1]=Y[2];
dY[0]=Y[1];
```

We will use a 4th and a 6th order method to solve this problem (Equations 2.12/RK4a and 2.21/RK6d). The code can be found in the on-line archive in file blasius.c. This has already been compiled and will run on any version of Windows®. A small batch file, _compile_blasius.bat, is also provided should you want to recompile it. The results can be found in blasius.xls and are listed below in abbreviated form. The Excel® spreadsheet contains an explicit and implict Euler solution as well.

BLASIUS

	4th Order Runge-Kutta			6th Order Runge-Kutta		
X	Y	dY/dX	dY^2/d^2X	Y	dY/dX	dY^2/d^2X
0.0	0.00000	0.00000	0.33206	0.00000	0.00000	0.33206
0.4	0.02656	0.13276	0.33147	0.02656	0.13276	0.33147
0.8	0.10611	0.26471	0.32739	0.10611	0.26471	0.32739
1.2	0.23795	0.39378	0.31659	0.23795	0.39378	0.31659
1.6	0.42032	0.51676	0.29666	0.42032	0.51676	0.29666
2.0	0.65002	0.62977	0.26675	0.65002	0.62977	0.26675
4.0	2.30575	0.95552	0.06423	2.30575	0.95552	0.06423
6.0	4.27962	0.99897	0.00240	4.27962	0.99897	0.00240
8.0	6.27921	1.00000	0.00001	6.27921	1.00000	0.00001
10.0	8.27921	1.00000	0.00000	8.27921	1.00000	0.00000
12.0	10.27921	1.00000	0.00000	10.27921	1.00000	0.00000

A More Complex Problem: Parallel Equations

The next problem we consider (a thermal plume discharged into a flowing river) is more complex because there are several equations to solve in parallel, including: conservation of mass, momentum, and energy, as well as width and position. There is also the conservation of salt if this is a brine plume or the receiving water is salty. The model is described in detail in my TWRA paper.[10]

There is considerably more input and output associated with this problem. There is the composition of the plume and the ambient: temperature, salinity, and velocity. The plume may also be discharged from a slot or a round jet and

[9] Summiya, P., Numerical Solution of Non Linear Differential Equation by Using Shooting Techniques, International Journal of Mathematics and Its Applications, Vol. 4, No. 1-A, pp. 93-100, 2016.

[10] Benton, D. J., "Development of a Two-Dimensional Plume Model for Positively and Negatively Buoyant Discharges into a Stratified Flowing Ambient," *Tennessee Water Resources Symposium*, 1989.

the equations must handle both cases. We also want to display the results graphically, so we will create all of the necessary files to accomplish this. The entire code, along with four sample input files (plume1.txt, plume2.txt, plume3.txt, and plume4.txt), can be found in the on-line archive in files plume2d.*. The variables are listed in the following table:

variable	units	description
α	-	entrainment coefficient
b	ft	width/diameter of the plume
depth	ft	depth
difd	ft	diffuser diameter
difl	ft	diffuser length
dify	ft	diffuser elevation (up from bottom)
dil	-	mixing ratio (dilution)
Froude	-	densimetric Froude number
Qr	ft^3/sec	river flow
Qdis	ft^3/sec	discharge flow
ρ	lbm/ft^3	density
S	-	salinity
T	°F	temperature
ang	°	angle of inclination from the horizontal
u	ft/s	horizontal velocity of the plume
v	ft/s	vertical velocity of the plume
w	ft/s	velocity along centerline $w=sqrt(u^2+v^2)$
x	ft	horizontal distance from diffuser ports
y	ft	vertical distance from diffuser ports
z	ft	distance along the centerline

The governing equations in symbolic form are:

p(1)=x	horizontal coordinate of centerline
p(2)=y	vertical coordinate of centerline
$p(3)=r*w*b^2*\pi/4$	round jet mass flux
p(3)=r*w*b	slot jet mass flux/unit diffuser length
p(4)=p(3)*u	horizontal momentum flux
p(5)=p(3)*v	vertical momentum flux
p(6)=p(3)*t	thermal energy flux
p(7)=p(3)*s	salt flux

The implementation is more lengthy too:

```
double ftr(double y)
```

```
{/* river temperature interpolation function */
int i;
if(y>=yr[0])
  return tr[0];
for(i=1;i<ntr;i++)
  if(y<=yr[i-1]&&y>=yr[i])
    return tr[i-1]+(y-yr[i-1])*(tr[i]-tr[i-1])/(yr[i]-
 yr[i-1]);
return tr[ntr-1];
}
double fur(double y)
{/* river velocity interpolation function */
int i;
if(nur<1)
  return urave;
if(y>=yr[0])
  return ur[0];
for(i=1;i<nur;i++)
  if(y<=yr[i-1]&&y>=yr[i])
    return ur[i-1]+(y-yr[i-1])*(ur[i]-ur[i-1])/(yr[i]-
 yr[i-1]);
return ur[nur-1];
}
double fsr(double y)
{/* river salinity interpolation function */
int i;
if(nsr<1)
  return srave;
if(y>=yr[0])
  return sr[0];
for(i=1;i<nsr;i++)
  if(y<=yr[i-1]&&y>=yr[i])
    return sr[i-1]+(y-yr[i-1])*(sr[i]-sr[i-1])/(yr[i]-
 yr[i-1]);
return sr[nsr-1];
}
void plume(double z,double*p,double*dp)
{/* the differentials */
double alpha,b,froude,ramb,rplm,samb,splm,tamb,tplm,
 uamb,uave,vave,vplm,y;
/* check for unstable plume */
if(p[2]<=0.)
  {
  dp[2]=0.;
  dp[3]=0.;
  dp[4]=0.;
  dp[5]=0.;
  dp[6]=0.;
  return;
```

```
  }
/* calculate plume velocities */
y=p[1];
if(round)
  {
  uave=p[3]/p[2]/1.33;
  vave=p[4]/p[2]/1.33;
  }
else
  {
  uave=p[3]/p[2]/1.43;
  vave=p[4]/p[2]/1.43;
  }
vplm=sqrt(uave*uave+vave*vave);
/* calculate plume trajectory angle */
dp[0]=uave/vplm;
dp[1]=vave/vplm;
/* calculate plume temperature,density,and width */
tplm=p[5]/p[2];
splm=p[6]/p[2];
rplm=rho(tplm,splm);
if(round)
  b=sqrt(4.*p[2]/rplm/vplm/M_PI);
else
  b=p[2]/rplm/vplm;
tamb=ftr(y);
uamb=fur(y);
samb=fsr(y);
ramb=rho(tamb,samb);
if(round)
  {
  froude=vplm/sqrt(fmax(0.000001,fabs(g*(ramb-
 rplm)*b/2./ramb)));
  if(acons>0.)
    alpha=acons;
  else
    alpha=0.0535*exp(1.43/sq(fmax(3.2,fmin(16.,
 froude))));
  dp[2]=M_PI*b*alpha*ramb*hypot(uamb-uave,vave);
  dp[3]=dp[2]*uamb;
  dp[4]=(ramb-rplm)*g*M_PI*b*b/4.;
  dp[5]=tamb*dp[2];
  dp[6]=samb*dp[2];
  }
else
  {
  froude=vplm/sqrt(fmax(0.000001,fabs(g*(ramb-
 rplm)*b/ramb)));
  if(acons>0.)
```

```
      alpha=acons;
    else
      alpha=0.0520*exp(1.62/pow(fmax(2.4,fmin(20.,
   froude)),1.5));
    dp[2]=alpha*ramb*hypot(uamb-uave,vave);
    dp[3]=dp[2]*uamb;
    dp[4]=(ramb-rplm)*g*b;
    dp[5]=tamb*dp[2];
    dp[6]=samb*dp[2];
    }
}
```

The system of equations is solved using 4th order Runge-Kutta (RK4c). Typical input is listed below:

```
test case#1 specified river flow/uniform velocity
50 46000 60000 depth,river cross-section,river flow
118.8 0 2500 700 17 .37 0 33 0 Tdis,Sdis,Qdis,difl,
   difd,dify,slotwidth,angle,alpha
10 0 0 number of temperatures, number of velocities
46.7 71.9 0 0 (z,t,v,s) if(nv<>0) (read in velocities)
45.1 71.4 0 0 if(nv=nt)then ignore velocities
43.4 70.4 0 0
42.6 69.4 0 0
39.3 68.9 0 0
36.0 68.5 0 0
26.2 67.9 0 0
16.3 67.5 0 0
 6.5 67.2 0 0
 0.0 67.0 0 0
```

Results for the first example (a rising plume) are shown in this figure:

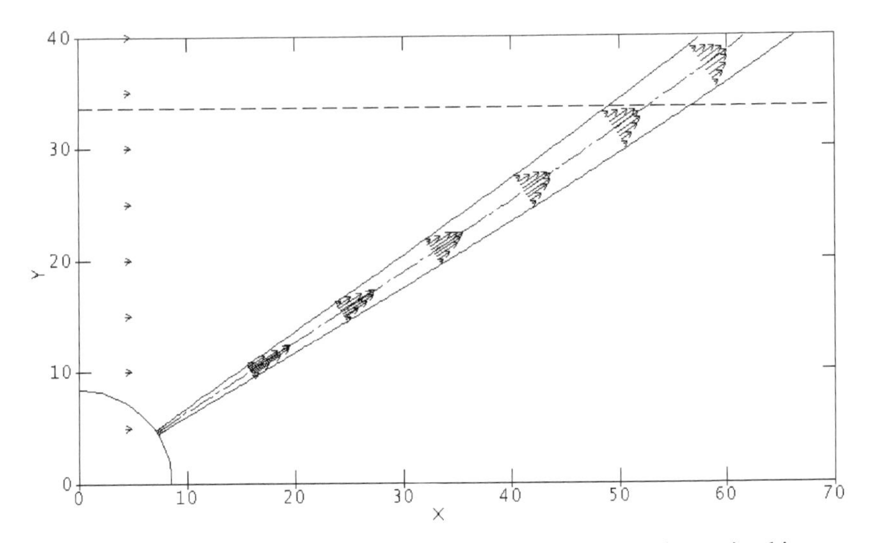

Results for the fourth example (a sinking plume) are shown in this next figure:

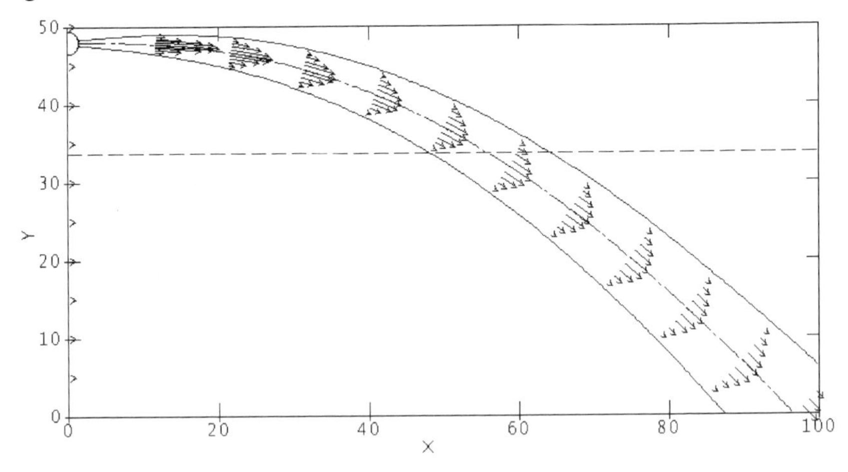

Crossflow Cooling Tower

The next problem we consider is combined heat and mass transfer within the packing inside a crossflow evaporative cooling tower. This next figure illustrates a crossflow cooling tower:

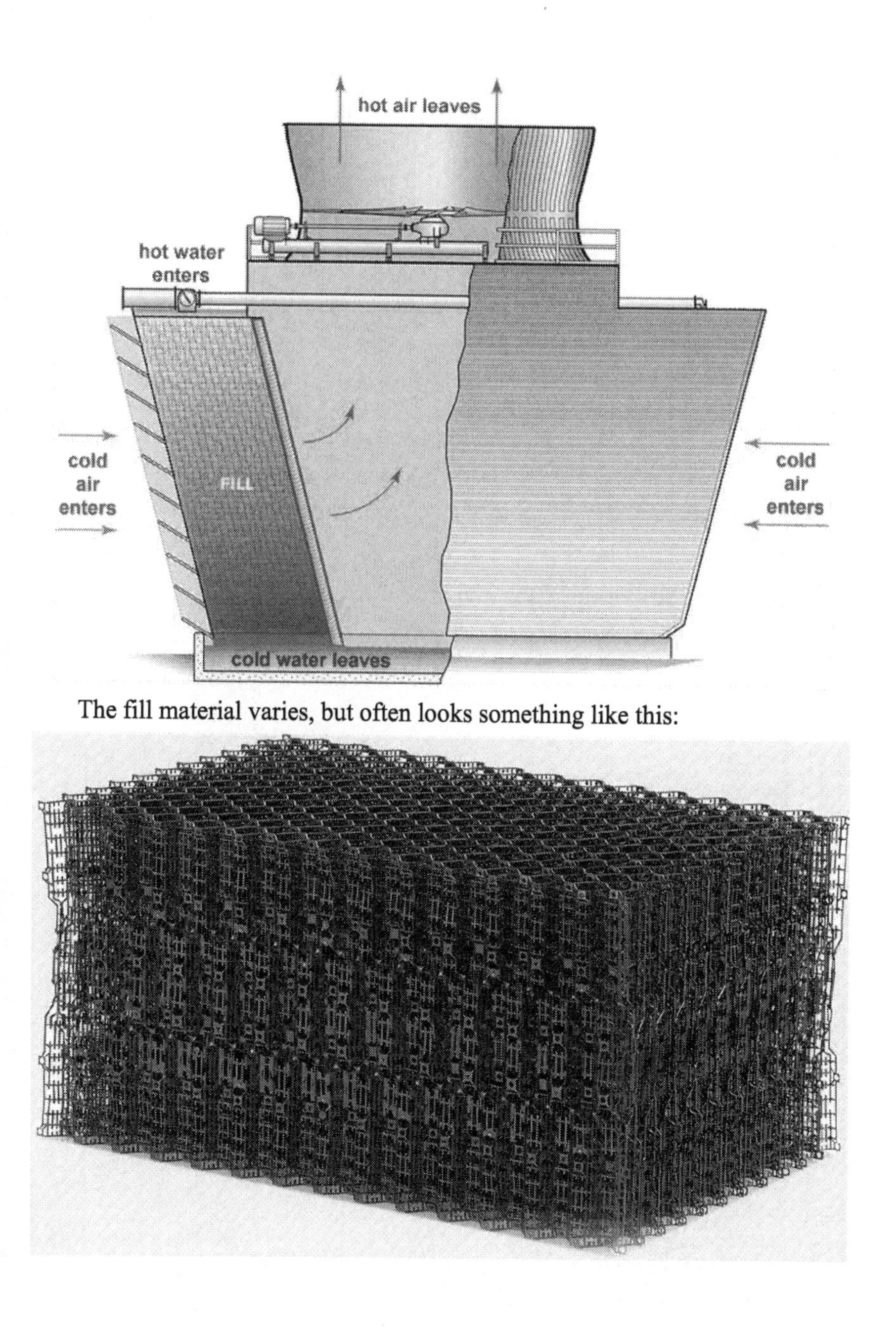

The fill material varies, but often looks something like this:

The governing equation goes back to Merkel.[11] In differential form the conservation of energy within a cell can be expressed by:

$$GKaY\left(h_W - h_A\right) - LCdT_W = 0 \tag{3.7}$$

In this equation G is the mass flux of air (lb/ft²/hr or kg/m²/s), a is the interfacial area per unit volume (ft²/ft³ or m²/m³), Y is the height of the packing (ft or m), h_W and h_A are the enthalpies of saturated moist air at the temperature of the water and air, respectively (BTU/lb or kJ/kg), L is the mass flux of the water (lb/ft²/hr or kg/m²/s), C is the specific heat of the water (BTU/lb/°F or kJ/kg/°C), and T_W is the temperature of the water (°F or °C). The factor K in Equation 3.7 is the dimensionless mass transfer coefficient. Merkel was interested in counterflow cooling towers and used the 4-point Chebyshev method to solve the equation. Here, we will consider crossflow. The differential is quite simple:

```
void Cell(double X,double Y,double Ha,double*dHa,double
   Tw,double*dTw)
  {
  double Hw,Q;
  Hw=fHtwb(user.baro,Tw);
  Q=user.KaY*(Hw-Ha);
  dHa[0]=Q*user.LG;
  dTw[0]=-Q;
  }
```

We start at the upper left hand corner of a rectangular block of cells. The hot water enters this cell from the top and the ambient air enters from the left. We can then use 4th order Runge-Kutta to integrate across and down, also very simple:

```
for(y=0;y<Ny;y++)
  {
  for(x=0;x<Nx;x++)
    {
    H=Ha[(Nx+1)*y+x];
    T=Tw[Nx*y+x];
    RungeKutta2D(Cell,X,1./Nx,Y,1./Ny,&H,&T);
    Ha[(Nx+1)*y+x+1]=H;
    Tw[Nx*(y+1)+x]=T;
    }
  }
```

The air and water temperatures (Ta and Tw) along with the air and water enthalpies (ha and hw) for a 5x5 grid are illustrated below:

[11] Merkel, F. Verdunstungskulung, V.D.I. Forschungsarbeiteh (Society of German Engineers Technical Journal), No. 275, Berlin, 1925.

						Tw				
Ta						148	148	148	148	148
78	102	116	124	128	132	115	115	124	130	131
78	92	99	108	115	118	97	104	108	113	121
78	86	92	98	103	108	88	96	99	104	110
78	82	88	92	96	101	84	89	93	98	102
78	80	84	87	91	94	81	85	89	92	97
						11 approach				89
						Hw				
Ha						260.1	260.1	260.1	260.1	260.1
41.6	74.7	107.4	131.5	149.2	165.8	104.7	105.8	132.7	157.3	162.5
41.6	59.3	70.8	86.9	103.7	114.1	66.8	79.0	87.5	100.9	122.9
41.6	50.5	58.7	68.0	77.6	88.2	53.6	64.4	69.2	79.0	93.4
41.6	46.0	52.8	58.6	64.5	73.4	48.0	54.5	59.9	68.0	74.5
41.6	44.0	48.0	51.8	57.5	62.2	45.2	49.3	54.6	59.1	66.2

This calculation is used to draw what are called *demand* curves. These tell to what value of K is required to the water to a certain temperature at a certain ratio of water to air flux. The packing (various sorts of fill material, most often plastic) provides some level of performance, which we call *supply*. The intersection of the supply and demand curves is where the cooling tower will operate. This is how cooling towers are designed.

We start with a value of entering air temperature (cool ambient, not necessarily uniform across the vertical face of the packing), entering water temperature (hot and uniform across the top of the fill), and a value of KaY/L. Then we step through the fill using Runge-Kutta to solve the differential equation, ultimately obtaining an average exiting (cooled) water temperature. We want curves of equal exiting water temperature, or more precisely, equal approach (exiting water temperature minus entering wet-bulb temperature). We adjust the value of KaY/L in order to match the desire approach and eventually draw a curve.

To accomplish this we use a bisection search. We pick a lower and upper limit on KaY/L and then iterate, each time cutting the interval in half until the result is within $1/2^{32}$ of the original span of KaY/L. This simple technique is very useful and works in many cases to solve nonlinear problems. Typical curves are:

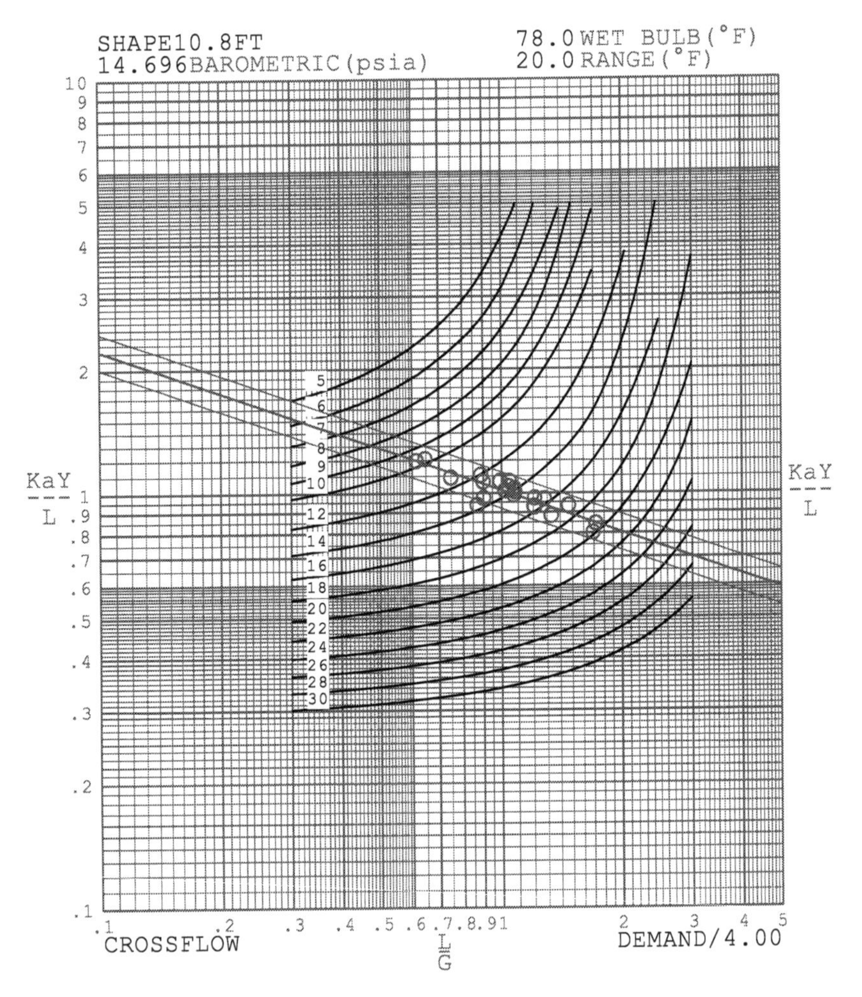

You will find a program (with source code) in the on-line archive in the folder examples\KaVL that is interactive, solves this problem for you, and creates graphs in English or SI units. The input dialog is:

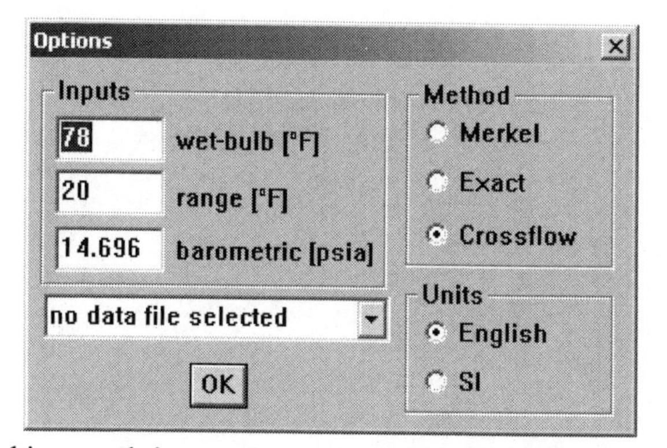

The resulting graph is pasted onto the clipboard. The code to produce the graph and copy it to the clipboard is also useful in other applications. Feel free to use it!

Chapter 4. Step Length Control

Considerable effort has been devoted to arriving at an error estimate for various Runge-Kutta methods so as to control the step length automatically. The standard approach is to calculate the result (i.e., the next step in the dependent variable) with two similar methods, one having a higher order than the other. We could simply use RK3a and RK4a or some other such combination; however, a desire to make this process more efficient has motivated theorists to find methods that use the same intermediate steps. For this reason these procedures are sometimes called *embedded* methods, although *parallel* methods might be more descriptive. We will consider three such methods. The first is called Huen-Euler and has the following Butcher tableau:

0	0	0
1	1	0
	1/2	1/2
	1	0

(4.1)

Instead of a single bottom row containing b_i, there are two such rows, one for each of the estimates. The second method is attributed to Fehlberg:

0	0	0	0
1/2	1/2	0	0
1	-1/256	255/256	0
	1/256	255/256	0
	1/512	255/256	1/512

(4.2)

The third is called Bogacki-Shampine:

0	0	0	0	0
1/2	1/2	0	0	0
3/4	0	3/4	0	0
1	2/9	1/3	4/9	0
	2/9	1/3	4/9	
	7/24	1/4	1/3	1/8

(4.3)

The following code implements each of these methods:

```
double HuenEuler(void dYdX(double,double*,
   double*),double*X,double dX,double*Y)
  {
```

```
  double k1,k2,W;
  dYdX(*X,Y,&k1);
  W=*Y+k1*dX;
  dYdX(*X+dX,&W,&k2);
  *Y=*Y+(k1+k2)*dX/2.;
  *X=*X+dX;
  return(*Y-W);
  }
double Fehlberg(void dYdX(double,double*,
   double*),double*X,double dX,double*Y)
  {
  double k1,k2,k3,W;
  dYdX(*X,Y,&k1);
  W=*Y+k1*dX/2.;
  dYdX(*X+dX/2.,&W,&k2);
  W=*Y+(k1+255.*k2)*dX/256.;
  dYdX(*X+dX,&W,&k3);
  *Y=*Y+(k1+510.*k2+k3)*dX/512.;
  *X=*X+dX;
  return(*Y-W);
  }
double BogackiShampine(void dYdX(double,double*,
   double*),double*X,double dX,double*Y)
  {
  double k1,k2,k3,k4,W;
  dYdX(*X,Y,&k1);
  W=*Y+k1*dX/2.;
  dYdX(*X+dX/2.,&W,&k2);
  W=*Y+3.*k2*dX/4.;
  dYdX(*X+3.*dX/4.,&W,&k3);
  W=*Y+(2.*k1+k2+4.*k3)*dX/9.;
  dYdX(*X+dX,&W,&k4);
  *Y=*Y+(7.*k1+6.*k2+8.*k3+3.*k4)*dX/24.;
  *X=*X+dX;
  return(*Y-W);
  }
double CashCarp(void dYdX(double,double*,
   double*),double*X,double dX,double*Y)
  {
  double k1,k2,k3,k4,k5,k6,W;
  dYdX(*X,Y,&k1);
  W=*Y+k1*dX/5.;
  dYdX(*X+dX/5.,&W,&k2);
  W=*Y+(3.*k1+9.*k2)*dX/40.;
  dYdX(*X+12.*dX/40.,&W,&k3);
  W=*Y+(3.*k1-9.*k2+12.*k3)*dX/10.;
  dYdX(*X+6.*dX/10.,&W,&k4);
  W=*Y+(-11.*k1+135.*k2-140.*k3+70.*k4)*dX/54.;
  dYdX(*X+dX,&W,&k5);
```

```
W=*Y+(3262.*k1+37800.*k2+4600.*k3
  +44275*k4+6831.*k5)*dX/110592.;
dYdX(*X+7.*dX/8.,&W,&k6);
W=*Y+(9361.*k1+38500.*k3+20125.*k4
  +27648.*k6)*dX/95634.;
*Y=*Y+(2808050.*k1+10550600.*k3+6721925.*k4
  +531009.*k5+6870528.*k6)*dX/27482112.;
*X=*X+dX;
return(*Y-W);
}
```

We will use the first example problem in Chapter 1 for a test case (Equation 1.1). The code can be found in the on-line archive in file embedded.c. The results can be found in embedded.xls. Since we know the exact solution (Equation 1.2), we can calculate the exact error and will just consider the absolute values. We solve the same problem for 4 different step sizes (h=0.05, 0.10, 0.15, and 0.20). The exact error (E2-E1 in the following code) is equal to the estimated change in the dependent variable (i.e., the result of applying the Runge-Kutta method for a single step, E2=Y2-Y1) minus the actual change in the dependent variable (i.e., the difference in the analytical solution for the same step in the independent variable, E1=f2-f1). We can write a single function to handle all four methods:

```
void testMethod(double method(void dYdX(double,double*,
    double*),double*,double,double*),double dX)
  {
  double dY1,dY2,E1,E2,f1,f2,X,Y1,Y2;
  X=Y2=f2=0.;
  do{
    f1=f2;
    Y1=Y2;
    E2=method(dYdX,&X,dX,&Y2);
    f2=f(X);
    dY1=f2-f1;
    dY2=Y2-Y1;
    E1=dY2-dY1;
    }while(X<3.);
  }
```

We can use the preceding function to test all four methods at four different step sizes:

```
for(dX=0.05;dX<0.21;dX+=0.05)
  {
  testMethod(HuenEuler,dX);
  testMethod(Fehlberg,dX);
  testMethod(BogackiShampine,dX);
  testMethod(CashCarp,dX);
  }
```

Results for the Huen-Euler method are shown in the following figure:

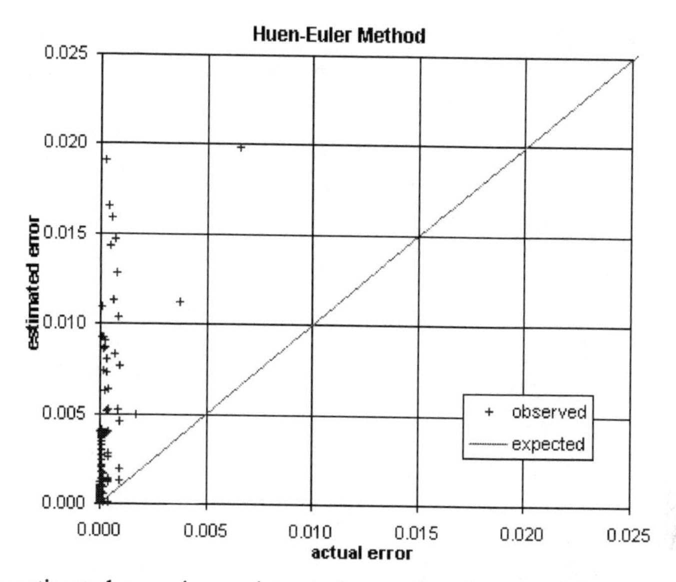

The estimated error is consistently larger than the actual (i.e., most of the blue +s are above the red line). There is also no pattern to the estimated error with respect to the actual error. Therefore, this method is useless—at least in this particular case. This next figure shows the same thing only for the Bogacki-Shampine method:

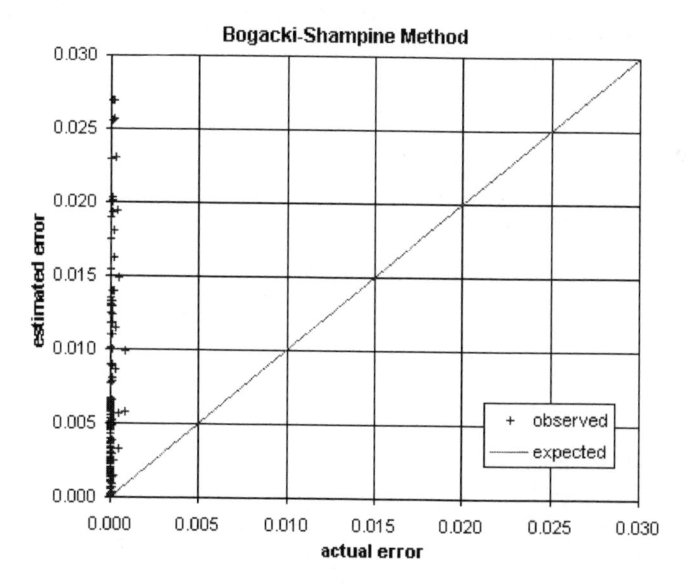

The estimated error is consistently larger than the actual (i.e., most of the blue +s are above the red line). Again, there is no pattern. This method is also useless—at least in this particular case. This next figure shows the Fehlberg:

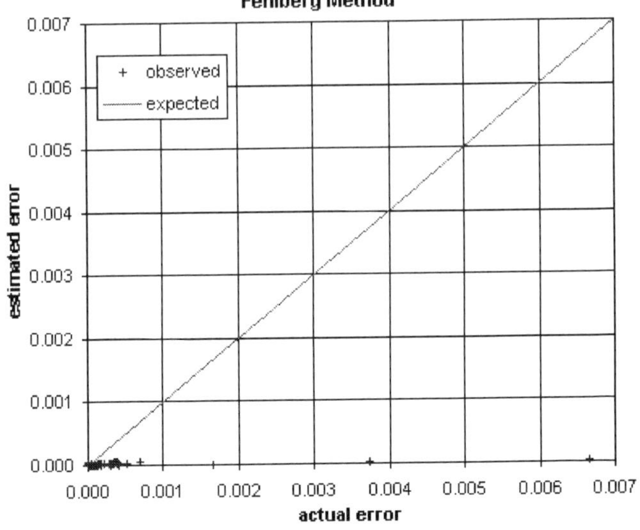

This time the estimated error is consistently smaller than the actual (i.e., most of the blue +s are below the red line). Still, there is also no pattern. This method is also useless—in this particular case. Next the Cash-Carp method:

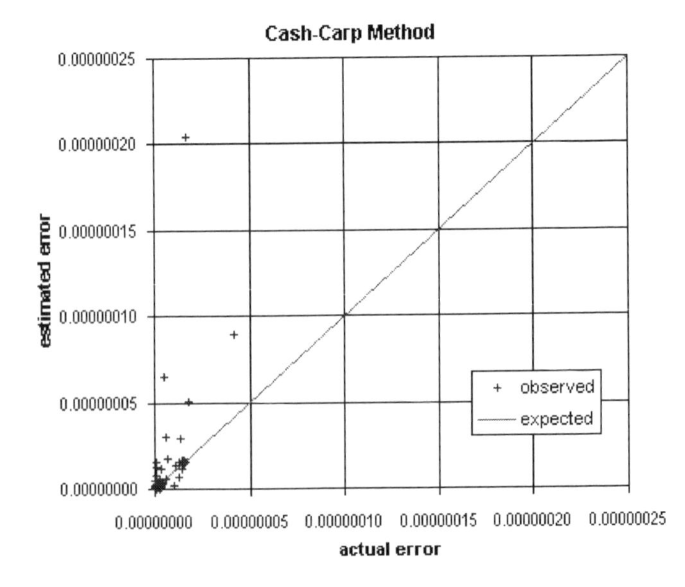

Finally, we start to see a pattern! An exploded view of the lower left corner of this figure shows the pattern in more detail:

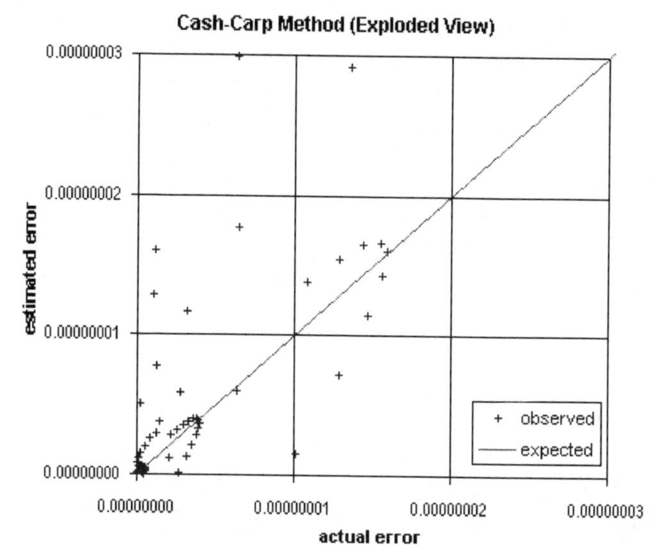

We could probably "fix" the first three methods, but what would be the point? This entire process has theoretical value, but is of little practical use. It might have been important before the invention of computers capable of performing billions of instructions per second, but that was so many nanoseconds ago that I've lost count.

Instead of applying a double method that yields an estimate of the error in twice as many microseconds, why not just cut the step size in half? If you don't get the same answer, cut it in half again. We've already presented code that calls a function, that calls another function, that implements whatever method from a table. Just put the whole thing inside a function and try smaller steps until it converges to your satisfaction.

Chapter 5. Finite Difference Method

Marching methods such as Runge-Kutta work well enough for initial value problems and some very simple boundary value problems, but they don't work for most boundary value problems, especially when the domain is complex. Finite difference methods work well for simple geometries. For complex geometries, finite element methods are preferable. We will cover those in Chapter 6.

Steady-State Heat Conduction

You rarely see the transient heat conduction equation with variable properties written out in its full form. In Cartesian coordinates (x,y,z) this is:

$$\frac{\partial(\rho CT)}{\partial t} = \frac{\partial}{\partial x}\left(k\frac{\partial T}{\partial x}\right) + \frac{\partial}{\partial y}\left(k\frac{\partial T}{\partial y}\right) + \frac{\partial}{\partial z}\left(k\frac{\partial T}{\partial z}\right) \tag{5.1}$$

Most often, the density (ρ), specific heat (C), and thermal conductivity (k) are presumed constant and taken outside the differential. In this first finite difference problem, we will consider steady-state (i.e., time-independent) conduction through a bar having variable thermal conductivity. In this case, Equation 5.1 reduces to:

$$0 = \frac{\partial}{\partial x}\left(k\frac{\partial T}{\partial x}\right) \tag{5.2}$$

As this is a second-order differential equation, we must consider at least three points, representing $x\text{-}\Delta x$, x, and $x+\Delta x$. We also need to know the thermal conductivity at $x\text{-}\Delta x/2$ and at $x+\Delta x/2$ in order to complete the expression:

$$0 = \frac{\left(k_{x+\frac{\Delta x}{2}}\frac{T_{x+\Delta x} - T_x}{\Delta x}\right) - \left(k_{x-\frac{\Delta x}{2}}\frac{T_x - T_{x-\Delta x}}{\Delta x}\right)}{\Delta x} \tag{5.3}$$

If Δx is constant, this reduces to:

$$k_{x+\frac{\Delta x}{2}}\left(T_{x+\Delta x} - T_x\right) = k_{x-\frac{\Delta x}{2}}\left(T_x - T_{x-\Delta x}\right) \tag{5.4}$$

Equation 5.4 can be solved for T_x:

$$T_x = \frac{k_{x+\frac{\Delta x}{2}}T_{x+\Delta x} + k_{x-\frac{\Delta x}{2}}T_{x-\Delta x}}{k_{x+\frac{\Delta x}{2}} + k_{x-\frac{\Delta x}{2}}} \tag{5.5}$$

We can easily construct an Excel® spreadsheet (finite_difference1.xls) to solve this problem. Because the temperatures depend on the thermal conductivities and the thermal conductivity depends on the temperature, this is an implicit (and nonlinear) calculation. Excel® can be programmed to handle this automatically (set tools options calculations iteration ON). For this example, we can define the thermal conductivity in a table (cells L3:M13) and use an interpolation macro (included) to find the value. We define the temperature at the end points (cells B2 and B12) and Excel® does the rest. We must help the iterations get started, so we use the IFERROR() function to assume an initial value. The result is:

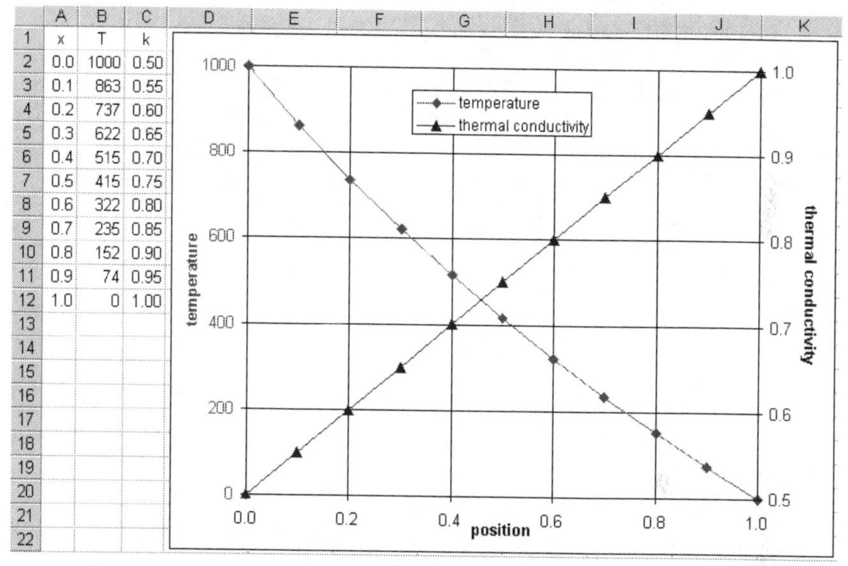

	A	B	C
1	x	T	k
2	0.0	1000	0.50
3	0.1	863	0.55
4	0.2	737	0.60
5	0.3	622	0.65
6	0.4	515	0.70
7	0.5	415	0.75
8	0.6	322	0.80
9	0.7	235	0.85
10	0.8	152	0.90
11	0.9	74	0.95
12	1.0	0	1.00

If the thermal conductivity were a constant, the blue curve would be flat and the red curve would be a straight line between the two end points. The formula in cell B3 is =IFERROR((B2*(C3+C4)+B4*(C2+C3))/(C2+2*C3+C4),500), which is Equation 5.5. We can extend this same finite difference to a second dimension, y, and solve for the corresponding result:

$$T_{xy} = \frac{\Delta y\left(k_{x+\frac{\Delta x}{2}} \mathrm{T}_{\mathrm{x+\Delta x}} + k_{x-\frac{\Delta x}{2}} \mathrm{T}_{\mathrm{x-\Delta x}}\right) + \Delta x\left(k_{y+\frac{\Delta y}{2}} \mathrm{T}_{\mathrm{y+\Delta y}} + k_{y-\frac{\Delta y}{2}} \mathrm{T}_{\mathrm{y-\Delta y}}\right)}{+\Delta y\left(k_{x+\frac{\Delta x}{2}} + k_{x-\frac{\Delta x}{2}}\right) + \Delta x\left(k_{y+\frac{\Delta y}{2}} + k_{y-\frac{\Delta y}{2}}\right)} \quad (5.6)$$

On the 2D tab of this same spreadsheet there is a two-dimensional problem with a rectangular domain having $\Delta x = \Delta y$ and prescribed temperatures along each of the 4 boundaries. It's also iterative and uses the same lookup table for thermal conductivity. You can change the temperatures along any of the boundaries or the thermal conductivities in the table and the solution will automatically adjust. Note that the corner cells in the temperature and thermal conductivity array are all empty, as these aren't used for anything and don't enter into the calculations. The result is:

	A	B	C	D	E	F	G	H	I	J	K	L	M
1							Temperatures						
2		1000	900	800	700	600	500	400	300	200	100	0	
3	1000	911	816	722	629	537	446	357	270	185	105	37	0
4	950	847	751	661	573	489	406	325	248	174	106	47	0
5	900	794	699	612	529	450	374	301	231	165	103	49	0
6	850	746	654	571	493	419	348	281	217	156	99	48	0
7	800	701	614	536	462	393	327	264	204	148	95	46	0
8	750	659	578	505	436	371	309	250	193	140	90	44	0
9	700	618	545	477	413	352	293	237	184	133	86	42	0
10	650	579	514	453	393	335	280	226	175	127	82	39	0
11	600	543	487	432	376	322	269	217	167	120	77	37	0
12	550	511	466	415	363	311	260	209	160	114	71	32	0
13	500	490	452	405	355	304	253	203	154	107	62	23	0
14		500	450	400	350	300	250	200	150	100	50	0	
15							Thermal Conductivities						
16		0.50	0.54	0.57	0.62	0.66	0.71	0.76	0.81	0.87	0.93	1.00	
17	0.50	0.53	0.57	0.61	0.65	0.69	0.73	0.78	0.83	0.88	0.93	0.98	1.00
18	0.52	0.56	0.59	0.63	0.67	0.71	0.75	0.80	0.84	0.89	0.93	0.97	1.00
19	0.54	0.58	0.62	0.65	0.69	0.73	0.77	0.81	0.85	0.89	0.93	0.97	1.00
20	0.56	0.60	0.64	0.67	0.71	0.75	0.79	0.82	0.86	0.90	0.93	0.97	1.00
21	0.57	0.62	0.65	0.69	0.73	0.76	0.80	0.83	0.87	0.90	0.94	0.97	1.00
22	0.59	0.63	0.67	0.70	0.74	0.77	0.81	0.84	0.87	0.91	0.94	0.97	1.00
23	0.62	0.65	0.69	0.72	0.75	0.78	0.82	0.85	0.88	0.91	0.94	0.97	1.00
24	0.64	0.67	0.70	0.73	0.76	0.79	0.82	0.86	0.89	0.92	0.95	0.97	1.00
25	0.66	0.69	0.71	0.74	0.77	0.80	0.83	0.86	0.89	0.92	0.95	0.98	1.00
26	0.68	0.70	0.72	0.75	0.78	0.81	0.84	0.87	0.90	0.92	0.95	0.98	1.00
27	0.71	0.71	0.73	0.76	0.78	0.81	0.84	0.87	0.90	0.93	0.96	0.98	1.00
28		0.71	0.73	0.76	0.79	0.81	0.84	0.87	0.90	0.93	0.97	1.00	

The temperatures in the rose cells are calculated using Equation 5.6 and the thermal conductivities in the sky blue cells are calculated from these using the interpolation macro. Once again, we must use the IFERROR() function to help Excel® get things started. If we were writing a code to do this, we would simply initialize all the variables to some reasonable value before starting to iterate for a solution.

Notice that cells A3 and B2 are equal, as are L2 and M3, A13 and B14, L14 and M13. These don't have to be equal in this spreadsheet, but in our finite difference scheme, these corner points represent the same location in space. In the real world, you don't have temperature discontinuities. To be realistic, these should be equal.

The above temperature map is illustrated below:

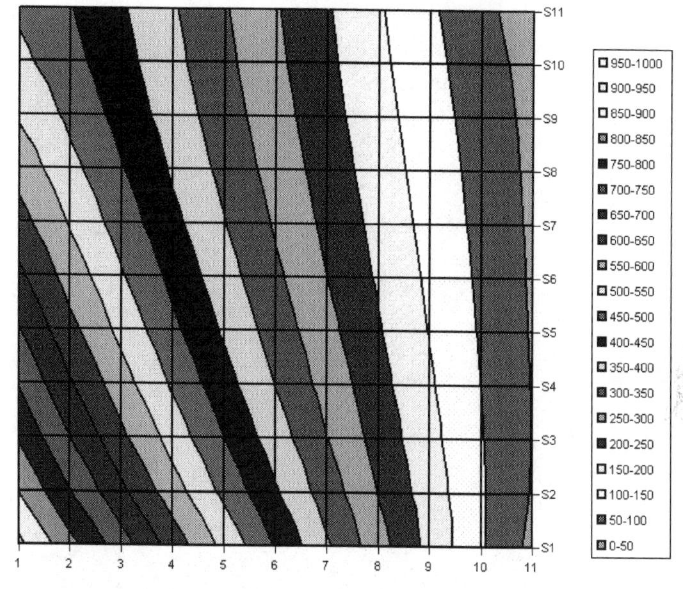

and here:

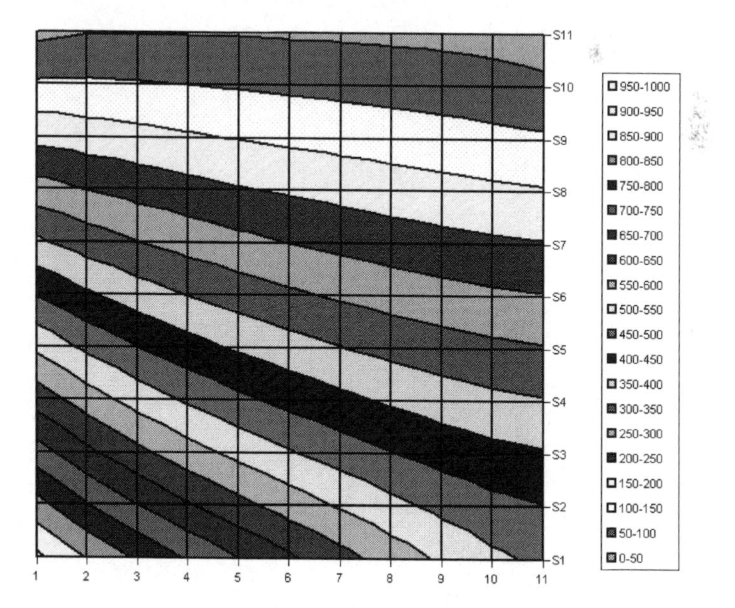

These are both sort of upside down and backwards because that's the way Excel® does it. The 2D graphs provided by Excel® leave much to be desired. To produce a quality 2D graph, you'll need a tool like Tecplot® or TP2. The latter is available free at the following link:

http://www.dudleybenton.altervista.org/software/index.html

TP2 produces the following proper rendering of this temperature field:

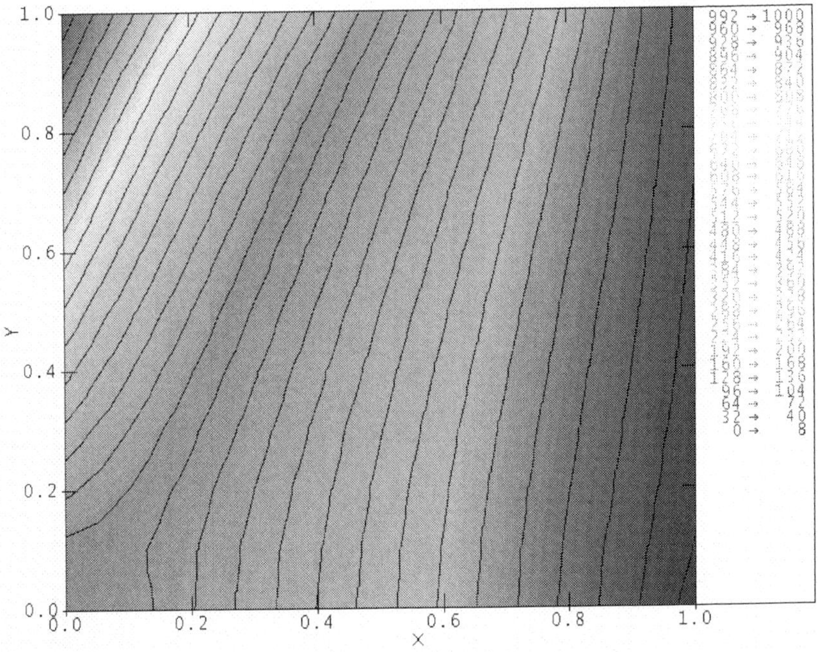

More details on finite difference operators may be found in Appendix B.

Transient Heat Conduction

No discussion of finite difference methods would be complete without mention of the Crank-Nicholson technique. For one-dimensional transient conduction with approximately constant properties Equation 5.1 simplifies to:

$$\rho C \frac{\partial T}{\partial t} = k \frac{\partial^2 T}{\partial x^2} \tag{5.7}$$

We might express this in terms of finite differences:

$$\rho C \left(\frac{T_{t+\Delta t,x} - T_{t,x}}{\Delta t} \right) = k \left(\frac{T_{t,x+\Delta x} - 2T_{t,x} + T_{t,x-\Delta x}}{\Delta x^2} \right) \tag{5.8}$$

Equation 5.8 is a central difference in space and a fully explicit (forward Euler) difference in time. It would be more accurate and also more stable to evaluate the right-hand side of Equation 5.8 at time t and also at $t+\Delta t$, using the average to compute the temporal derivative. This would be an implicit difference in time, often called the *midpoint* method. This requires knowledge of the temperatures before and after the time step in order to calculate each step through time.

If we had planned to solve a matrix for the temperatures at the next step, this merely adds a few more terms, which can be rearrange and folded into the calculations we were already going to perform. This is the Crank-Nicholson method and is quite stable, even for large time steps. While this idea is very clever and quite effective, it has little practical use for several reasons. First, you rarely want to use the kind of algorithms required to solve this type of matrix. Second, Crank-Nicholson is fairly easy to encode for a 1D problem, but it's a real hassle for 2D and beyond. Who needs to solve 1D transient heat conduction problems? Third, you can just march through time using a Runge-Kutta method and forget about solving a matrix.

Because the Crank-Nicholson method is interesting theoretically and historically significant, we will set up and solve a simple problem in one dimension with 4 nodes. The 4 nodal point equations become:

$$\begin{bmatrix} 1+\beta & -\beta/2 & 0 & 0 \\ -\beta/2 & 1+\beta & -\beta/2 & 0 \\ 0 & -\beta/2 & 1+\beta & -\beta/2 \\ 0 & 0 & -\beta & 1+\beta \end{bmatrix} \times \begin{bmatrix} T_{1,t+\Delta t} \\ T_{2,t+\Delta t} \\ T_{3,t+\Delta t} \\ T_{4,t+\Delta t} \end{bmatrix} = \begin{bmatrix} 1-\beta & \beta/2 & 0 & 0 \\ \beta/2 & 1-\beta & \beta/2 & 0 \\ 0 & \beta/2 & 1-\beta & \beta/2 \\ 0 & 0 & \beta & 1-\beta \end{bmatrix} \times \begin{bmatrix} T_{1,t} \\ T_{2,t} \\ T_{3,t} \\ T_{4,t} \end{bmatrix} \qquad (5.9)$$

The parameter β is equal to:

$$\beta = \frac{k\Delta t}{\rho C \Delta x^2} \qquad (5.10)$$

The entire process can be implemented in an Excel® spreadsheet (Crank-Nicholson.xls in the on-line archive) using the matrix multiply MMULT() and matrix inversion MINVERSE() functions. The matrices and properties are:

Crank-Nicholson Matrices								properties	
matrix A				matrix B				Δt	1.00
2.0	-0.5	0.0	0.0	0.0	0.5	0.0	0.0	Δx	1.00
-0.5	2.0	-0.5	0.0	0.5	0.0	0.5	0.0	ρ	1.00
0.0	-0.5	2.0	-0.5	0.0	0.5	0.0	0.5	C	1.00
0.0	0.0	-1.0	2.0	0.0	0.0	1.0	0.0	k	1.00
calculated parameter, β=kΔt/(ρCΔx²)									1.00

Change the numbers in bold blue and the results will automatically update.

The temperatures are shown in this next figure:

Crank-Nicholson time steps								
calculated temperatures at each time step								
t=0	t=Δt	t=2Δt	t=3Δt	t=4Δt	t=5Δt	t=6Δt	t=7Δt	t=8Δt
100	46.4	38.0	30.8	26.3	22.5	19.3	16.5	14.2
100	85.6	66.3	56.7	48.3	41.5	35.6	30.6	26.2
100	95.9	85.1	73.2	63.1	54.1	46.5	39.9	34.3
100	97.9	90.5	79.1	68.1	58.6	50.3	43.2	37.1

The initial temperatures are in the first column (all set to 100). Each time step to the right is an application of the matrix operations in Equation 5.9. The same thing could be implemented with a marching technique, the simplest being:

explicit marching scheme								
0	0	0	0	0	0	0	0	0
100	75.0	56.3	42.2	31.6	23.7	17.8	13.3	10.0
100	100.0	87.5	78.1	66.9	57.1	47.9	40.0	33.0
100	100.0	100.0	91.7	82.6	72.2	62.1	52.7	44.2
100	100.0	95.8	87.2	77.4	67.2	57.4	48.4	40.5
d²/dx²								
-50	-37.5	-28.1	-21.1	-15.8	-11.9	-8.9	-6.7	-5.0
0	-25.0	-18.8	-22.4	-19.6	-18.4	-15.9	-13.9	-11.8
0	0.0	-16.7	-18.1	-20.9	-20.0	-18.9	-16.9	-14.9
0	-8.33	-17.4	-19.5	-20.5	-19.5	-17.9	-15.9	-13.9

Open Channel Flow

The conservation of mass and momentum for unsteady flow in an open channel is described by the following two partial differential equations, respectively:

$$\frac{\partial A}{\partial t} + \frac{\partial Q}{\partial x} + q = 0 \tag{5.11}$$

$$\frac{\partial Q}{\partial t} + gA\frac{\partial h}{\partial x} + \frac{1}{A}\frac{\partial Q^2}{\partial x} = gA\left(S_0 - S_f\right) \tag{5.12}$$

Here, Q is the flow (or discharge, as it is often called in open channel flow), A is the cross sectional area, q is the lateral inflow (not contributing momentum along the channel), g is the acceleration of gravity, h is the depth (or water surface elevation), x is the distance along the channel, t is time, S_f is the friction slope, and S_0 is the bottom slope (positive with decline in the downstream direction). Equation 5.12 is called the De-Saint Venant Equation.

The slope (geometric and effective arising from friction along the channel bottom) for open channel flow is empirically described by Manning's Equation:

$$Q = VA = \left(\frac{1.49}{n}\right) AR^{\frac{2}{3}} \sqrt{S} \tag{5.13}$$

In the preceding equation V is the bulk velocity, A is the cross-sectional area, R is the hydraulic radius, S is the geometric slope, and n is Manning's factor. The factor 1.49 arises from the unit conversion for meters to feet $3.2808^{1/3}$. This is unity when using SI units.

We will solve the conservation of mass and momentum equations simultaneously using a finite difference approach. The more interesting aspect of this example is that of sloshing, which is something water often does, especially when you open one dam and shut another. Sloshing makes this a *stiff* system of equations, meaning the derivatives with respect to time can easily run away with overshoot.

We will use two techniques to handle the problem of sloshing: 1) predictor/corrector and 2) MacCormack spatial derivatives. The former is like iteratively approximating the midpoint method discussed in the previous example. The latter has an effect of reducing sloshing like *numerical viscosity*.

The simple predictor corrector method is easily implemented by initializing the new values of flow and elevation equal to the old and then successively substituting revised values several times before advancing time to the next step. In this case experience has shown that 4 iterations is adequate. Partial differentials for the finite difference equations are calculated by the following function:

```
void Differentials(double*Q,double*H,double*dQdt,
   double*dHdt)
  {
  int i;
  double Bavg,dHdx,dQdx,dQQAdx;
  for(i=0;i<sections;i++)
    QQA[i]=Q[i]*Q[i]/A[i];
  for(i=0;i<sections-1;i++)
    {
    dQdx=(Q[i+1]-Q[i])/dX;
    Bavg=(B[i]+B[i+1])/2.;
    dHdt[i]=(Qinf[i]-dQdx)/Bavg;
    }
  for(i=1;i<sections-1;i++)
    {
    dHdx=(H[i]-H[i-1])/dX;
```

```
        dQQAdx=(QQA[i+1]-QQA[i-1])/2./dX;
        dQdt[i]=-gravity*A[i]*(dHdx+Sf[i])-dQQAdx;
        }
```

The step forward in time is accomplished by the following code with *ncorr* corrector iterations:

```
/* corrector: iterate ncor times */
           for(iter=0;iter<ncorr;iter++)
             {
/* update slopes */
             for(i=1;i<sections-1;i++)
               Sf[i]=So[i]*Q[new][i]*fabs(Q[new][i]);
/* calculate differentials (pass new arrays) */
             Differentials(&Q[new][0],&H[new][0],
   &dQdt[new][0],&dHdt[new][0]);
/* step forward in time */
             for(i=0;i<sections-1;i++)
               {
               H[new][i]=H[old][i]+dt*
    (dHdt[old][i]+dHdt[new][i])/2.;
               H[new][i]=max(Hmin,min(Hmax,H[new][i]));
               }
             for(i=1;i<sections-1;i++)
               Q[new][i]=Q[old][i]+dt*
    (dQdt[old][i]+dQdt[new][i])/2.;
             }
           hour+=1./nstep;
```

The predictor/corrector step is initialized by the following code:

```
/* predictor: set Hnew and Qnew to Hold and Qold */
           for(i=0;i<sections-1;i++)
             {
             H[new][i]=H[old][i]+dt*dHdt[old][i]/2.;
             H[new][i]=max(Hmin,min(Hmax,H[new][i]));
             }
           for(i=1;i<sections-1;i++)
             Q[new][i]=Q[old][i]+dt*dQdt[old][i]/2.;
```

The MacCormack upwind/downwind differencing scheme is accomplished by storing the elevation, H, and flow, Q, in a two-dimensional arrays:

```
double H[2][sections-1];/* water surface elevation */
double dHdt[2][sections-1];
double dQdt[2][sections];
double Q[2][sections]; /* flow */
```

and the rest of the variables in one-dimensional arrays:

```
double A[sections]; /* area */
double B[sections]; /* width */
double Havg[sections]; /* average elevation */
double Qinf[sections-1]; /* local inflow */
```

```
double QQA[sections]; /* Q²/A */
double Sf[sections-1]; /* friction slope */
double So[sections]; /* total slope */
```

The spatial finite differences are flip-flopped every time step by swapping the integer index *new* and *old*.

```
/* swap new/old index for MacCormack step */
new=1-new;
old=1-new;
```

The differencing function is called with the index of the arrays so that this changes with each step, but doesn't require any additional code to implement beyond this:

```
/* calculate differentials (pass new arrays) */
Differentials(&Q[new][0],&H[new][0],&dQdt[new][0],
   &dHdt[new][0]);
```

The function receives a pointer to the first element in either the new or old section of the array, respectively:

```
void Differentials(double*Q,double*H,double*dQdt,
   double*dHdt)
```

In order to get the calculations started, two initialization days are run with the same inputs before the actual calculations:

```
  if(initialize)
    {
/* first time through take two days to initialize
   everything */
    ndays=3;
    hour=-24.*(ndays-1);
    for(i=0;i<sections-1;i++)
      H[new][i]=hdn+(hup-hdn)*pow((sections-2.-
   i)/(sections-2.),4);
    for(i=0;i<sections;i++)
      Q[new][i]=Qups[0]+(Qdns[0]-Qups[0])*i/(sections-
   1.);
    }
else
  initialize=FALSE;
```

The runoff (drainage into the reservoir from rain) and local inflow from a side stream are added before the iteration begins:

```
/* add runoff to each cell */
    for(i=0;i<sections-1;i++)
      Qinf[i]=Runoff;
/* add inflow from stream */
    Qinf[infsection]+=Qstream/dX;
```

The geometry (area, width, and slope) are initialized using the predictor values:

```
    Havg[0]=H[new][0];
    for(i=1;i<sections;i++)
      Havg[i]=(H[new][i-1]+H[new][i])/2.;
    Geometry(Havg,A,B,So);
```

The input is supplied in 24-hour blocks, read from a file. Typical input is:

```
680 680 0 0 <-- WATTS BAR TAILWATER ELEV./CHICKAMAUGA
   HEADWATER ELEV./
 1      0     0              TOTAL LOCAL INFLOW/HIWASSEE
   RIVER INFLOW
 2      0     0
 3      0     0  <-- HOUR/WATTS BAR, CHICKAMAUGA RELEASE
   CFS
 4 10000 10000
 5 10000 10000
 6 15000 10000
 7  5000 20000
 8  5000 20000
 9  5000 30000
10 15000 30000
11 20000 30000
12 30000 30000
13 30000 30000
14 40000 30000
15 40000 30000
16 40000 30000
17 40000 30000
18 40000 30000
19 40000 30000
20 20000     0
21 10000     0
22     0     0
23     0     0
24     0     0
```

This particular geometry is for the Chickamauga Reservoir on the Tennessee River between Watts Bar Dam and Chickamauga Dam. The side stream entering the channel is the Hiawassee River. The midsection where flow and elevation are of special interest is at Soddy Daisy, the site of Sequoyah Nuclear Plant. All of the associated files can be found in the on-line archive named SQFLOW.*. The following is typical output of the model:

```
Chickamauga One-Dimensional Reservoir Routing Model
using multi-step predictor-corrector method
initializing day 1
initializing day 2
routing day 1
hour  Qups   Qmid  Qdns  Hups   Hmid   Hdns
 100     0 -14567     0 680.33 680.12 680.18
 200     0 -15484     0 680.16 680.01 680.00
```

300	0	3186	0	680.23	679.98	679.87
400	10000	18744	10000	680.73	679.99	679.97
500	10000	17710	10000	680.80	680.02	680.04
600	15000	8011	10000	680.92	680.05	680.05
700	5000	16007	20000	680.14	680.06	679.98
800	5000	22859	20000	679.81	679.99	679.99
900	5000	21634	30000	679.96	679.97	679.89
1000	15000	22959	30000	680.83	679.85	679.83
1100	20000	17727	30000	681.28	679.77	679.73
1200	30000	23909	30000	682.12	679.70	679.62
1300	30000	32524	30000	682.49	679.69	679.60
1400	40000	32546	30000	683.54	679.69	679.66
1500	40000	29620	30000	684.06	679.71	679.65
1600	40000	30575	30000	684.31	679.71	679.66
1700	40000	31543	30000	684.45	679.73	679.66
1800	40000	33243	30000	684.54	679.75	679.69
1900	40000	32784	30000	684.61	679.77	679.72
2000	20000	22324	0	682.64	679.91	679.91
2100	10000	4075	0	680.90	680.12	680.18
2200	0	5655	0	679.61	680.18	680.23
2300	0	7815	0	679.39	680.22	680.21
2400	0	-1130	0	679.94	680.23	680.26

A graph of the results is shown in this next figure:

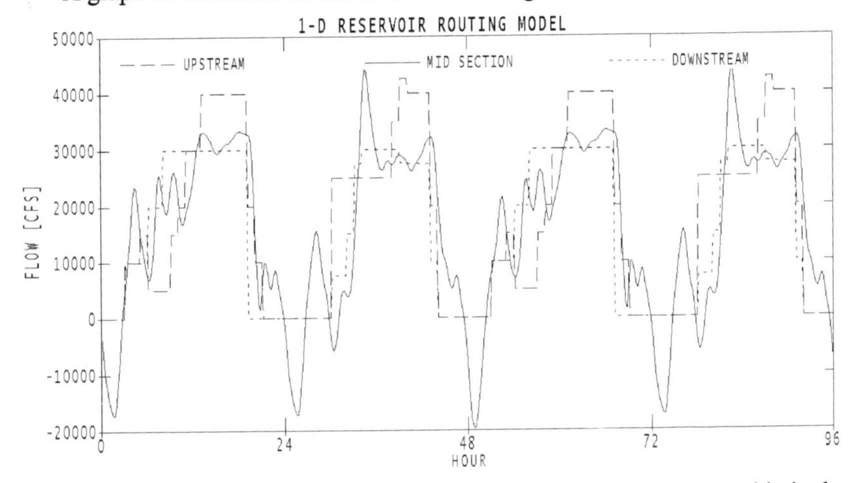

The upstream and downstream flows have a step shape, because this is the way the dams are operated: gates are either open or closed and there are a fixed number of open settings for each gate. Flow at the midpoint exhibits significant sloshing. In fact, the sloshing is so pronounced that 49 hours into this cycle, the flow is back upstream at a rate of 20,000 cubic feet per second (566m³/s).

When you close the downstream dam, it sends a wave back up stream. The midsection Sequoyah shown by the solid line is 12.5 miles (20.1 km) upstream of Watts Bar dam. These *waves* travel all the back to the upstream dam 58.6 miles (94.3 km) and are reflected back, resulting in smaller highs and lows in the flow at subsequent times.

Open Channel Flow with Heat Transfer and Dry Bed

The preceding model will only work with a fully-wetted, mostly full channel. It also doesn't consider thermal energy or temperature. We will now modify the equations to handle the case when all the water runs out, leaving the bed dry. We will also handle heat transfer so that the temperature may be calculated. In the preceding model, both the upstream and downstream were regulated by dams. In this next case, the upstream will be controlled by a dam, but the downstream is free to discharge into a larger body of water, so that it has no rigid boundary condition.

The seminal reference on kinematic waves is the USGS monograph by Miller.[12] This work (which is readily available on-line) explains the development of the governing equations as well as various solutions and has been the principal guidance for surface water models ever since. The kinematic wave equation is a special case of Equation 5.11.

$$\frac{1}{c}\frac{\partial Q}{\partial t}+\frac{\partial Q}{\partial x}=0 \qquad (5.14)$$

The friction term (i.e., right-hand side of Equation 5.11) has disappeared, as momentum is presumed dominant. The new term, c, is the celerity (i.e., wave propagation speed). As it turns out, we will need to add the friction term back in order to better fit observed data. At this point of Miller's development, the celerity is rather vague, perhaps equal to dQ/dA, but more likely $c=\sqrt{gh}$.

After a discussion of the method of characteristics, wave propagation and reflection, Miller presents the following finite difference equation:

$$Q_{x+\Delta x,t+\Delta t}=C_1 Q_{x,t}+C_2 Q_{x,t}+\Delta t+C_3 Q_{x+\Delta x,t} \qquad (5.15)$$

Equation 5.15 derives from the Muskingum-Cung method using variable parameters C_1, C_2, and C_3 as described by Ponce and Yevjevich.[13] The parameters are:

[12] Miller, J. E., "Basic Concepts of Kinematic-Wave Models," U. S. Geological Survey Professional Paper No. 1302, 1984.
[13] Ponce, V. M. and Yevjevch, V., "Muskingum-Cunge Method with Variable Parameters," *Journal of the Hydraulics Division* of the American Society of Civil Engineers, Vol. 104, HY 3, pp. 353-360, 1978.

$$C_1 = \frac{1+C-D}{1+C+D}$$
$$C_2 = \frac{-1+C+D}{1+C+D} \tag{5.16}$$
$$C_3 = \frac{1-C+D}{1+C+D}$$

In Equation 5.16 the parameter C is the Courant number and D is a dimensionless term that brings the friction factor back into play.

$$C = \frac{c\Delta t}{\Delta x}$$
$$D = \frac{q}{S_0 c \Delta x} \tag{5.17}$$

As with the previous routing model, we must initialize the calculations in order to get the process started:

```
/* initialize flows and celerrities */
  if(initialize)
    {
    initialize=FALSE;
    qinitl=400.;
    dx=12.5*5280./(sections-1);

   v=qinitl/B[0]/pow(qinitl*rn[0]/(B[0]*pow(So[0],0.5)*1
   .486),0.6);
    hnew=qinitl/v/B[0];
    for(i=0;i<sections;i++)
      {
      Q[new][i]=qinitl;
      H[new][i]=hnew;
      C[new][i]=1.5*v;
      Triv[new][i]=Tdam;
      Tbed[new][i]=Tdam;
      rvbed[i]=dbed*rhobed*B[i]*dx;
      }
    }
```

Then we swap the old and new index and iterate (predictor/corrector) to calculate the celerity because it's implicit and not simply equal to $\sqrt{gh}$.

```
/* switch old/new index */
  while(TRUE)
    {
    new=1-new;
    old=1-new;
/* approximate new values by old values */
```

```
      for(i=0;i<sections;i++)
        {
        Q[new][i]=Q[old][i];
        H[new][i]=H[old][i];
        C[new][i]=C[old][i];
        Triv[new][i]=Triv[old][i];
        }
/* determine maximum celerity */
      iter1=0;
      cmx=0.;
      while(TRUE)
        {
        iter1+=1;
        if(iter1>niter1)
          goto l_100;
        cmax=C[new][0];
        for(i=0;i<sections;i++)
          cmax=max(cmax,C[new][i]);
        Q[new][0]=Qdam;
        Triv[new][0]=Tdam;
/* if initial flow is not zero,check it with next
   boundary condition. whichever is greater must be used
   to calculate to retain stability */
        if(Q[new][0]>0.)
          C[new][0]=1.5*Q[new][0]/(B[0]*
   (Q[new][0]*rn[0]/(B[0]*pow(So[0],0.5)*1.486))*0.6);
        cmax=max(cmax,C[new][0]);
```

We must limit the step size in order to not exceed the Courant condition (i.e., disturbances can't propagate faster than $\Delta x/\Delta t$). We iterate for flow after the time step (predictor/corrector method again):

```
/* calculate step size and put an upper limit on dt */
        dt=min(dt0,dx/cmax);
/* start of loop that calculates the q for each section
   of the river */
        for(i=1;i<sections;i++)
          {
          iter2=0;
          while(TRUE)
            {
            iter2+=1;
/* check for non-convergence in flow */
            if(iter2>niter2)
              goto l_130;
            qbar=(Q[old][i-1]+Q[old][i])/2.;
            cbar=(C[new][i-1]+C[new][i]+C[old][i-
   1]+C[old][i])*0.25;
            qb=(Q[new][i-1]+Q[new][i])/2.;
            qbxt=(qbar+qb)/2.;
```

```
          theta=max(0.,min(0.9999,(1.-
    qbxt/(B[i]*So[i]*cbar*dx))/2.));
          c11=dx/(cbar*dt);
          c3=exp(-cbar*dt/(dx*(1.-theta)));
          c1=1.-c11*(1.-c3);
          c2=c11*(1.-c3)-c3;
          qnew=Q[new][i];
          Q[new][i]=c1*Q[new][i-1]+c2*Q[old][i-
    1]+c3*Q[old][i];
          if(Q[new][i-1]>Q[new][i]&&Q[old][i]>Q[new][i])
            Q[new][i]=Q[old][i];
          change=fabs(qnew-Q[new][i]);
          if(iter2>6)
            Q[new][i]=(Q[new][i]+qnew)/2.;
          if(iter2>1&&change<50.)
            break;
          }
        }
```

Then update the geometry, slopes, and depths, recalculate the friction factor, and correct the celerity—all of this inside an iteration loop.

```
  /* calculate new slopes and depths */
      for(i=0;i<sections;i++)
        {
        v=0.1;
        if(Q[new][i]>0.)
          {
  /* s is the friction slope term. put a lower limit on it
     so that the stage calculation will not have division
     by zero or sqrt of a negative number */
          s=max(0.000125,(Q[new][i]-
     Q[old][i])/(dt*B[i]*sq(C[new][i])+So[i]));
  /* this is the stage rating curve */
          H[new][i]=pow(Q[new][i]*rn[i]
     /(1.486*B[i]*sqrt(s)),0.6);
          }
  /* calculate velocity,celerity,flow */
        v=Q[new][i]/(B[i]*H[new][i]);
        C[new][i]=1.5*v;
        }
      change=fabs(cmax-cmx);
      if(iter1>1&&change<0.2)
        break;
      cmx=cmax;
      }
```

We calculate the temperature by the conservation of energy. The rate of increase in energy of the water in a cell is equal to that carried in by the water entering the cell minus that carried out by the water exiting the cell plus the heat transfer from the atmosphere to the water in the cell plus the heat transfer from the streambed into the water in the cell. The energy of any element of water is equal to the volume times the density times the heat capacity times the temperature ($E=V\rho CT$).

The rate of heat exchange with the air or streambed is equal to the area times the heat transfer coefficient times the temperature difference (heat transfer=$Ah\Delta T$). The symbols can become a bit confused here, as Q is often used to denote flow and heat transfer and h is often used to denote height and heat transfer coefficient. The calculation of temperature begins thus:

```
/* determine water temperature by conservation of energy
    (first law of thermodynamics) */
     for(i=1;i<sections;i++)
       {
       rvold=rho*H[old][i]*B[i]*dx;
       rvnew=rho*(Q[old][i-1]+Q[new][i-1]-Q[old][i]-
    Q[new][i])*dt/2.+rvold;
/* determine if bed is dry */
       drybed=FALSE;
       if(rvnew<1000.)
         drybed=TRUE;
/* determine energy in a section */
       ein=rho*dt*(Q[old][i-1]*Triv[old][i-1]+Q[new][i-
    1]*Triv[new][i-1])/2.;

    eout=rho*dt*(Q[old][i]*Triv[old][i]+Q[new][i]*Triv[ne
    w][i])/2.;
       eold=Triv[old][i]*rvold;
/* set bounds on temperature to assure agreement with
    the second law of thermodynamics */
       tmin=min(Triv[old][i-1],Triv[old][i]);
       tmin=min(tmin,Triv[new][i-1]);
       tmin=min(tmin,T0);
       tmin=min(tmin,Tbed[old][i]);
       tmax=max(Triv[old][i-1],Triv[old][i]);
       tmax=max(tmax,Triv[new][i-1]);
       tmax=max(tmax,T0);
       tmax=max(tmax,Tbed[old][i]);
/* iterate to determine new temperature */
       iter3=0;
       while(TRUE)
         {
         iter3+=1;
/* check for non-convergence */
         if(iter3>niter3)
           goto l_120;
```

There is no advection, only convection, when the streambed is dry, so we must handle both cases:

```
/* skip river temperature calculations if bed is dry */
      change=0.;
      if(!drybed)
        {
/* calculate net thermal advection */
        thad=edifft*(B[i-1]*H[old][i-1]
   *(Triv[old][i-1]-Triv[old][i])+B[i]*H[old][i]
   *(Triv[old][min(sections,i+1)-1]-Triv[old][i])
   +B[i-1]*H[new][i-1]*(Triv[new][i-1]-Triv[new][i])
   +B[i]*H[new][i]*(Triv[old][min(sections,i+1)-1]
   -Triv[new][i]))/2./dx;
/* calculate surface heat transfer */
        surfht=dx*B[i]*htcsur*(2.*T0-Triv[new][i]
   -Triv[old][i])/2./3600.;
        }
/* estimate the heat transfer coefficient between the
   river and the bed in a section and then calculate
   heat transfer to the river bed */
      if(drybed)
        {
        htcbed=htcsur;
        rbedht=dx*B[i]*htcbed
   *(Tbed[new][i]+Tbed[old][i]-2.*T0)/2./3600.;
        }
      else
        {
        htcbed=2./(H[new][i]/kwater+(dbed/kbed));
        rbedht=dx*B[i]*htcbed
   *(Tbed[new][i]+Tbed[old][i]-Triv[new][i]
   -Triv[old][i])/2./3600.;
        }
/* calculate new energy in a section of the river */
      enew=eold+ein-eout+(surfht+thad+rbedht)*dt;
      if(!drybed)
        tnew=enew/rvnew;
      tnew=max(tmin,min(tmax,tnew));
      change=fabs(Triv[new][i]-tnew);
      Triv[new][i]=(Triv[new][i]+tnew)/2.;
/* calculate new river bed temperature in a section */
      tnew=Tbed[old][i]-rbedht*dt/rvbed[i]/cbed;
      tnew=max(tmin,min(tmax,tnew));
      change=max(change,fabs(Tbed[new][i]-tnew));
      Tbed[new][i]=(Tbed[new][i]+tnew)/2.;
      if(drybed)
        Triv[new][i]=Tbed[new][i];
/* check for convergence in temperature */
      if(iter3>1&&change<1.)
        break;
      }
```

```
        }
```

We then update time and check for convergence of the predictor/corrector step.

```
/* increment time */
     time+=dt;
/* check to see if time step process has caught-up */
     if(time>=dt0)
       goto l_110;
     }
/* convergence error in celerity */
l_100:
  return(1);
```

The geometry for this model is based on the Clinch River, which flows out of Norris Dam and downstream 12.5 miles (20.1 km), where it enters Watts Bar Reservoir. Originally, the stream could run shallow during times of low releases from the dam, but small rock-and-cage impoundments have been added to prevent this from happening and regulate the aquatic habitat. As with the previous model, an input file is read and results written to an output file for plotting. All of the associated files can be found in the on-line archive named INFLOW.*. Typical results are:

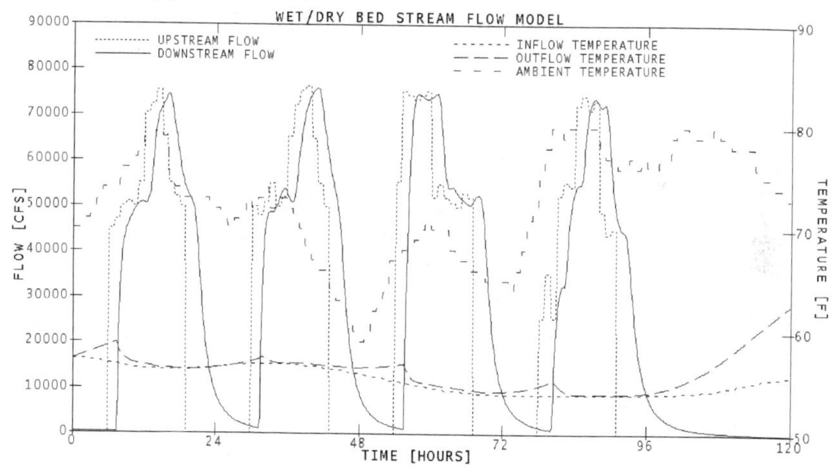

There is no sloshing in this river, as there is no downstream dam to abruptly cut off the flow and send waves back upstream.

Chapter 6. Finite Element Method

Sadly, the finite element method is more often than not presented with pomp and circumstance in a dense obscuring fog with much mathematical mumbling and a great deal of hand waving. All this is to impress the student with the wonder of it all and the breath-taking knowledge of the lecturer. I hate to break with tradition, but I'm not going to do that.

The Ensemble Hypothesis

Before we begin exploring the finite element method, we must introduce the *Ensemble Hypothesis*. In short, the ensemble hypothesis states that the whole may be greater than the sum of the individual parts. As it applies to model development, we propose that an ordered assemblage of distinguishable parts may adequately approximate the whole, in spite of the fact that no single part displays all of the characteristics of the whole.

For example, we may break an irregular domain into a collection of triangular or tetrahedral elements. We may know from experience that the distribution of the dependent variable (temperature, concentration, stress, strain, etc.) does not vary linearly over the domain, but it may over a single element. There are also boundaries distinct from the interior of the domain. We may have one type of element along the boundaries and a different type within the interior. We can put it all together in such a way that the ensemble adequately represents the whole.

Optimizing the Whole vs. the Parts

The finite difference method focuses on representing the governing equation within a single element in terms of differences that approximate the calculus. The resulting algebraic equations relate the values of the dependent variable at the nodal points, based on the differentials across each element. Various direct and indirect methods are used to solve the algebraic equations, but the approach for obtaining the equations is the same. The finite difference approximations are selected to optimally approximate the differentials. This is focusing on the *parts*.

The finite element approach is different. It starts with an unknown *basis* function that approximates the dependent variable throughout a single element. The basis function may be as simple as $T(x,y)=a+bx+cy$. The constants (a, b, and c) may be related to the temperatures at the three corners of a triangle so that the basis function matches at these three nodes. Instead of trying to minimize the error in approximating a differential within each element, we want to minimize the error in approximating the governing equation over the entire domain. We express this error as an integral over the area or volume that is the domain.

Calculus of Variations

This topic always comes up when introducing the finite element method and you may have wondered why. The reason is simple: it's the branch of mathematics in which we discuss the extrema (i.e., minima or maxima) of such things as integrals. You have no doubt seen derivations for the minimum or maximum of a function where you simply take the derivative and set it to zero. When the function is an integral over some domain, it's not so simple.

To find the optimum of an ensemble of things, you can't simply remove the integral and set what's inside to zero. It doesn't work that way and it's easy to come up with examples to illustrate this. For instance, if you want to maximize the profit from an amusement park, you must consider many things. If you raise the price of each ticket, fewer people may come. If you lower the ticket price, more people may come, but you might lose money in the process. You also can't simply charge the first rich kid who comes through a million dollars and let everyone else in free.

When it comes optimizing integrals, there are rules—new rules you may not have seen before advanced calculus, where this topic is discussed. The rules are logical and can be properly derived and also demonstrated by example. Here's the secret to the finite element method: We must find a function, such that when we integrate it over the domain and then apply the rules to arrive at the minimum error, we end up with the equation we're trying to solve in the first place. There's an expression for this process: *finding the corresponding variational problem*. Consider the following integral:

$$\iiint_{Domain} \left\{ \left(\frac{\partial T}{\partial x} \right)^2 + \left(\frac{\partial T}{\partial y} \right)^2 + \left(\frac{\partial T}{\partial y} \right)^2 \right\} dV \Rightarrow \min \tag{6.1}$$

After we apply the rules of variational calculus to Equation 6.1, to find the minimum, we arrive at:

$$\iiint_{Domain} \left\{ \frac{\partial^2 T}{\partial x^2} + \frac{\partial^2 T}{\partial y^2} + \frac{\partial^2 T}{\partial z^2} \right\} dV = 0 \tag{6.2}$$

Inside the integral is Laplace's equation.

$$\frac{\partial^2 T}{\partial x^2} + \frac{\partial^2 T}{\partial y^2} + \frac{\partial^2 T}{\partial z^2} = 0 \tag{6.3}$$

If we start with a basis function that satisfies Equation 6.1 and solve it over the domain, applying the various boundary conditions, it *will* approximate Laplace's equation with the minimum residual over the entire domain. Fortunately, Laplace's equation governs a lot of problems, including: heat transfer, diffusion, magnetic and electrostatic fields , stress, and strain. Sadly, nobody has ever found the corresponding variational problem for the Navier-Stokes equations (i.e., the equations of fluid flow). Still, there are many different approximations and we will discuss some of these, but leave solutions to another book.

2D Conduction Model

We have already introduced a basis function that satisfies Equation 6.1 and that is T(x,y)=a+bx+cy. We will use this basis function to solve several problems governed by Laplace's equation. There are a lot of programming details necessary to read in and check a finite element model. Rather than interrupt the presentation of the finite element method at this point, these details are discussed in Appendix C. Here, we will jump into setting up and solving the equations. The nodal point equations are handled in a single concise function:

```
void NodalPointEquations()
  {
  int i,j,n1,n2,n3;
  double AKXY,A11,A12,A13,A22,A23,A33,GA3,
    X1,X12,X13,X2,X23,X3,XY,Y1,Y12,Y13,Y2,Y23,Y3;
  for(i=0;i<Ne;i++) /* element conservation equations */
    {
    n1=Ie[3*i]; /* node 1,2,3 from element Ie[]*/
    n2=Ie[3*i+1];
    n3=Ie[3*i+2];
    X1=Xn[n1]; /* x,y for each node from node[] */
    Y1=Yn[n1];
    X2=Xn[n2];
    Y2=Yn[n2];
    X3=Xn[n3];
    Y3=Yn[n3];
    for(j=0;j<Na[n1];j++) /* index I,J,K from Ia[] */
      Ip[Ia[Ma*n1+j]]=j;
    for(j=0;j<Na[n2];j++)
      Jp[Ia[Ma*n2+j]]=j;
    for(j=0;j<Na[n3];j++)
      Kp[Ia[Ma*n3+j]]=j;
    X12=X2-X1; /* inverse of T(x,y)=a+b*x+c*y */
    X13=X3-X1;
    X23=X3-X2;
    Y12=Y2-Y1;
    Y13=Y3-Y1;
    Y23=Y3-Y2;
    XY=X12*Y23-X23*Y12;
    AKXY=Ce[i]*Ae[i]/(XY*XY); /* k*area */
    A11= (X23*X23+Y23*Y23)*AKXY;
    A12=-(X13*X23+Y13*Y23)*AKXY;
    A13= (X12*X23+Y12*Y23)*AKXY;
    A22= (X13*X13+Y13*Y13)*AKXY;
```

```
    A23=-(X12*X13+Y12*Y13)*AKXY;
    A33= (X12*X12+Y12*Y12)*AKXY;
    Dn[n1]+=A11;
    Dn[n2]+=A22;
    Dn[n3]+=A33;
    An[Ma*n1+Ip[n2]]+=A12; /* left-hand side of A[] */
    An[Ma*n1+Ip[n3]]+=A13;
    An[Ma*n2+Jp[n1]]+=A12;
    An[Ma*n2+Jp[n3]]+=A23;
    An[Ma*n3+Kp[n1]]+=A13;
    An[Ma*n3+Kp[n2]]+=A23;
    GA3=Ge[i]*Ae[i]/3; /* generation term */
    Bn[n1]+=GA3; /* right-hand side of A[]X[]=B[] */
    Bn[n2]+=GA3;
    Bn[n3]+=GA3;
    }
  }
```

The core of this section of code is the inverse of the following matrix:

$$\begin{bmatrix} 1 & x_1 & y_1 \\ 1 & x_2 & y_2 \\ 1 & x_3 & y_3 \end{bmatrix} \quad (6.4)$$

There are three types of boundary conditions: isothermal (*T*=...), heat flux (*q*=...), and convection (*q=h*(T-To)*). These are added in the following simple function:

```
void BoundaryConditions()
  {
  int i,j,j1,j2,k,n,n1,n2;
  double b1,b2,hs,hs3,hs6,hst2,q,s;
  for(n=0;n<Nb;n++)
    {
    i=Ib[n];
    j1=Jb[n];
    j2=(j1+1)%3;
    n1=Ie[3*i+j1];
    n2=Ie[3*i+j2];
    k=Kb[n];
    b1=B1[n];
    b2=B2[n];
    s=Se[3*i+j1];
    if(k==0) /* isothermal */
      {
      Dn[n1]=1;
      Bn[n1]=b1;
      memset(An+Ma*n1,0,Ma*sizeof(double));
      Dn[n2]=1;
```

```
      Bn[n2]=b1;
      memset(An+Ma*n2,0,Ma*sizeof(double));
      }
    else if(k==1) /* heat flux */
      {
      q=b1*s/2;
      Bn[n1]+=q;
      Bn[n2]+=q;
      }
    else /* convection */
      {
      for(j=0;j<Na[n1];j++)
        Ip[Ia[Ma*n1+j]]=j;
      for(j=0;j<Na[n2];j++)
        Jp[Ia[Ma*n2+j]]=j;
      hs=b1*s;
      hs3=hs/3;
      hs6=hs/6;
      hst2=hs*b2/2;
      Bn[n1]+=hst2;
      Bn[n2]+=hst2;
      Dn[n1]+=hs3;
      Dn[n2]+=hs3;
      An[Ma*n1+Ip[n2]]+=hs6;
      An[Ma*n2+Jp[n1]]+=hs6;
      }
    }
  }
```

That's it. The rest is solving a sparse matrix. There are several methods available to accomplish this. The method used in this case is Successive Over Relaxation (SOR). This and several other methods are discussed in Appendix D. All of the files associated with this example can be found in the on-line archive in a folder named examples\finite elements\2D-conduction named con2d.*. The following is typical input:

```
LOWER HALF OF A SYMMETRIC CONVECTING FIN IN AN ISO-
   THERMAL SUBSTRATA
NE NN NB IPRT MAP NITER  RELAX  DTMAX
65 49 31   1   1   100    1.5    0.1
ELEMENT SPECIFICATIONS
 1  1  2  3 0 .2
 2  2  4  3 0 .2
 3  3 44  5 0 .2
 4  4  6 44 0 .2
etc...
BOUNDARY SPECIFICATIONS
 1  1 1 2 0 0
 2  2 1 2 0 0
 3  4 1 2 0 0
```

```
  4  6 1 2 0 0
 etc...
 NODAL POINT LOCATIONS
  1 .00 .08
  2 .02 .08
  3 .02 .07
  4 .04 .08
```

In order to adequately visualize the results, an advanced graphics program (such as TP2) is necessary. The program (CON2D) creates all of the necessary output files so that TP2 can display the results. Three examples are provided: fin, star, and tree. The fin below is what you might find on the head or cylinder of a motorcycle engine, dissipating heat to the air passing over it.

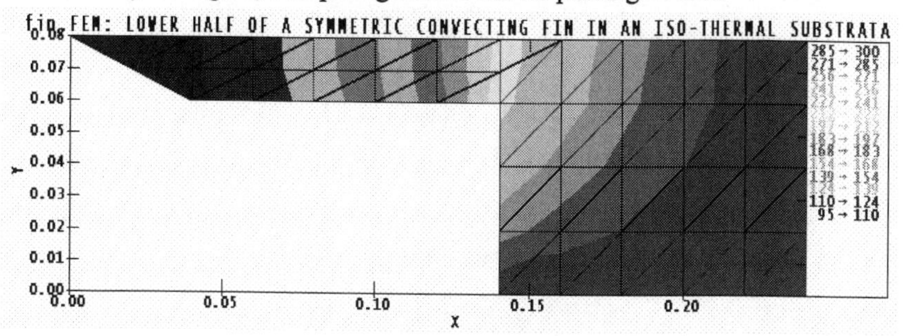

This next problem has internal heat generation at the center of the red zone:

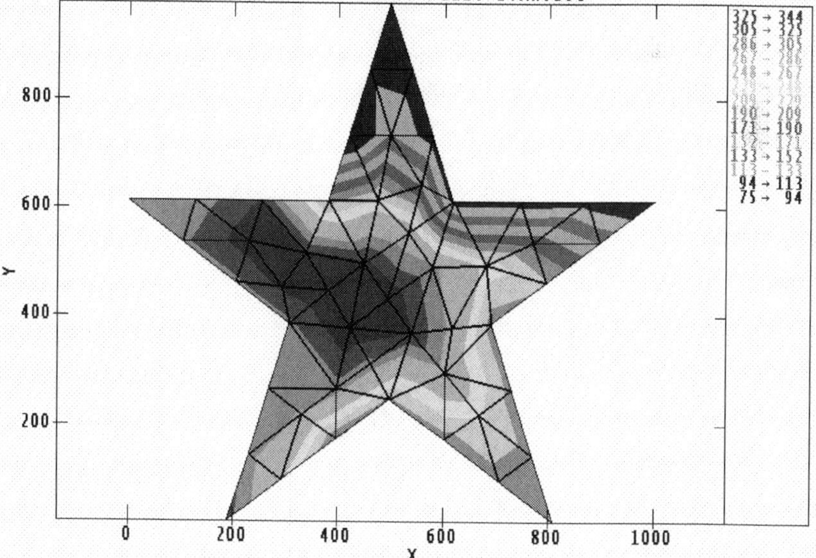

This third problem is just a three-pronged fin. There is a batch file to solve and plot each one of these examples (e.g., _solve_tree.bat). This is the results for the third problem:

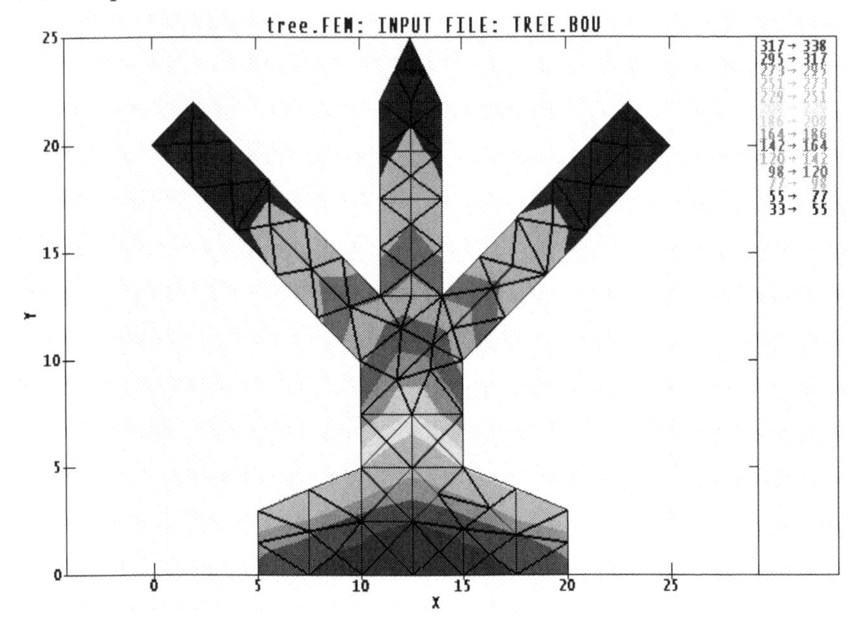

2D Plane Stress and Strain

The next problem we will consider is two-dimensional (x,y plane) stress and strain. The following figure illustrates plane stress:

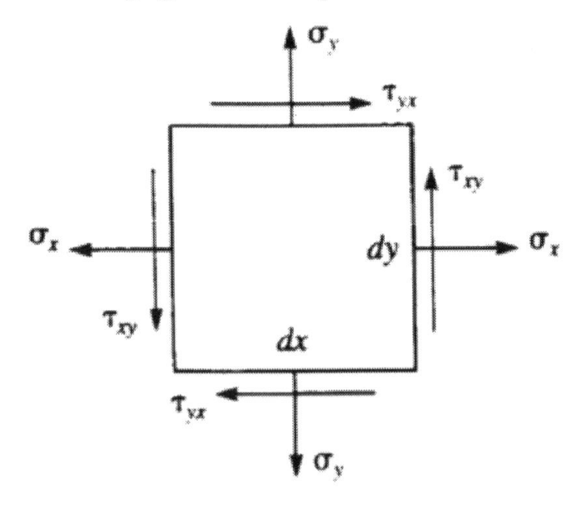

In two dimensions (x,y) the stress tensor (vector) has three parts:

$$\begin{bmatrix} \sigma_x & \sigma_y & \tau_{xy} \end{bmatrix}^T \tag{6.5}$$

Plane strains are illustrated in this next figure:

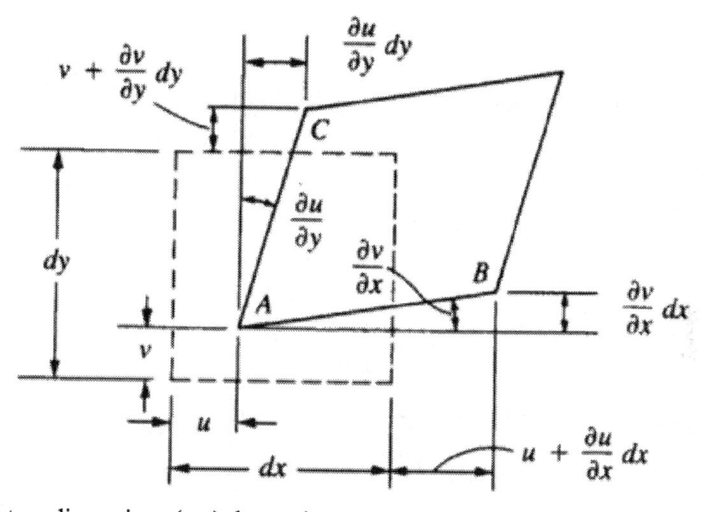

In two dimensions (x,y) the strain tensor (vector) also has three parts:

$$\left[\varepsilon_x = \frac{\partial u}{\partial x} \quad \varepsilon_y = \frac{\partial v}{\partial y} \quad \gamma_{xy} = \frac{\partial u}{\partial y} + \frac{\partial v}{\partial x} \right]^T \tag{6.6}$$

The stress and strain are related by Hooke's Law, which can be expressed in vector form:

$$\begin{Bmatrix} \sigma_x \\ \sigma_y \\ \tau_{xy} \end{Bmatrix} = [D] \begin{Bmatrix} \varepsilon_x \\ \varepsilon_y \\ \gamma_{xy} \end{Bmatrix} \tag{6.7}$$

For plane stress the matrix [D] is given by:

$$[D] = \frac{E}{1-\nu^2} \begin{bmatrix} 1 & \nu & 0 \\ \nu & 1 & 0 \\ 0 & 0 & \frac{1-\nu}{2} \end{bmatrix} \tag{6.8}$$

where E is Young's modulus and ν is Poisson's Ratio. For plane strain the matrix [D] is given by:

$$[D] = \frac{E}{(1+\nu)(1-2\nu)} \begin{bmatrix} 1-\nu & \nu & 0 \\ \nu & 1-\nu & 0 \\ 0 & 0 & \frac{1}{2}-\nu \end{bmatrix} \tag{6.9}$$

Because the stress and strain satisfy Laplace's equation, we use the same basis function, $f(x,y)=a+bx+cy$. In this section, we will only consider triangular elements, as you can build whatever you want out of them and there are various programs available to discretize most any shape. The area of a triangle is given by the determinant of the following 3x3 matrix:

$$A = \frac{1}{2} \begin{bmatrix} 1 & x_1 & y_1 \\ 1 & x_2 & y_2 \\ 1 & x_3 & y_3 \end{bmatrix} \tag{6.10}$$

Next, we define a basis vector containing the basis function at the three nodes of a triangular element:

$$[N] = \frac{1}{2A} \begin{bmatrix} a_1 + b_1 x + c_1 y \\ a_2 + b_2 x + c_2 y \\ a_3 + b_3 x + c_3 y \end{bmatrix} \tag{6.11}$$

Following Equation 6.2, we take the partial derivatives of the basis vector and integrate over the volume of an element to arrive at the following, known as the stiffness matrix, [K]:

$$[K] = V[B]^T [D][B] \tag{6.12}$$

where V is the volume, equal to the area, A, times the thickness, t, and [B] is the matrix resulting from the differentiation of [N]:

$$[B] = \frac{1}{2A} \begin{bmatrix} y_{23} & 0 & y_{31} & 0 & y_{12} & 0 \\ 0 & x_{32} & 0 & x_{13} & 0 & x_{21} \\ x_{32} & y_{23} & x_{13} & y_{31} & x_{21} & y_{12} \end{bmatrix} \tag{6.13}$$

where $x_{ij}=x_i-x_j$ and $y_{ij}=y_i-y_j$. These equations are simple. Most of the effort implementing the method is bookkeeping. All of the files associated with this section can be found in the on-line archive in the folder examples\finite element\plane strain. We begin with a typical input file:

```
8 6 1 2 nodes elements pointloads bearings
0 0 x y
100 0
100 100
0 100
200 0
200 100
300 0
300 100
1 2 3 200000 0.3 0.5 i j k Young Poisson thickness
1 3 4 200000 0.3 0.5
2 5 6 200000 0.3 0.5
2 6 3 200000 0.3 0.5
5 7 8 200000 0.3 0.5
5 8 6 200000 0.3 0.5
8 0 -10000 n Fx Fy
1 1 1 n Rx Ry
4 1 0
```

The number of nodes, elements, point loads, and bearings (constraints) are on the first line. After that we have x,y for each node on sequential lines until all are specified. Following that, we have the three nodes that define each element, along with Young's modulus, Poisson's Ratio, and the thickness for each element. This allows us to handle composites in which the properties vary from element-to-element. It doesn't matter what the units are, as long as they are consistent (length, force, force/length, and force/length^2).

After the nodes and elements, the point loads (force per unit thickness). These are specified by the number of the node to which the force is applied, along with the x and y components. The bearings (horizontal and/or lateral constraints) are defined in similar manner: node, x constraint, y constraint. A value of 0 indicates the point is not constrained and a value of 1 indicates that it is constrained.

This code (psfem.c) wants the elements to be defined in counter-clockwise orientation. To facilitate input, we perform a simple test. If the area of the element is negative, we swap the order of the elements. That way, it doesn't matter how you order the nodes in the input file.

```
double Area(int i,int j,int k)
  {
  double x1,x2,x3,y1,y2,y3;
  x1=Node[i].x;
  x2=Node[j].x;
  x3=Node[k].x;
  y1=Node[i].y;
  y2=Node[j].y;
  y3=Node[k].y;
  return(((x2-x1)*(y3-y2)-(x3-x2)*(y2-y1))/2.);
  }
```

```
    if(Area(i,j,k)>0.)
      {
      Elem[l].i=i;
      Elem[l].j=j;
      Elem[l].k=k;
      }
    else
      {
      Elem[l].i=i;
      Elem[l].j=k;
      Elem[l].k=j;
      }
    Elem[l].ym=ym;
    Elem[l].nu=nu;
    Elem[l].tk=tk;
    }
```

After reading the input file, we assemble the global matrix:

```
  for(i=0;i<elems;i++)
    {
    stif=Stiff(Node[Elem[i].i].x,Node[Elem[i].i].y,
               Node[Elem[i].j].x,Node[Elem[i].j].y,
               Node[Elem[i].k].x,Node[Elem[i].k].y);
    getKe(stif,Elem[i].ym,Elem[i].nu,Elem[i].tk,ke);
    if(!Test(ke))
      Abort(__LINE__,"element %i is degenerate",i+1);
    Global(ke,i);
    }
```

Within this loop, we calculate the stiffness matrix [K] for each element:

```
STIF Stiff(double x1,double y1,double x2,double
   y2,double x3,double y3)
  {
  static STIF stif;
  stif.x1=x1;
  stif.y1=y1;
  stif.x2=x2;
  stif.y2=y2;
  stif.x3=x3;
  stif.y3=y3;
  stif.x13=x1-x3;
  stif.x32=x3-x2;
  stif.x21=x2-x1;
  stif.x23=x2-x3;
  stif.y13=y1-y3;
  stif.y23=y2-y3;
  stif.y31=y3-y1;
  stif.y12=y1-y2;
  return(stif);
  }
```

```
void getB(STIF stif,double*B)
  {
  double det;
  det=stif.x13*stif.y23-stif.x23*stif.y13;
  B[ 0]=stif.y23/det;
  B[ 1]=0.;
  B[ 2]=stif.y31/det;
  B[ 3]=0.;
  B[ 4]=stif.y12/det;
  B[ 5]=0.;
  B[ 6]=0.;
  B[ 7]=stif.x32/det;
  B[ 8]=0.;
  B[ 9]=stif.x13/det;
  B[10]=0.;
  B[11]=stif.x21/det;
  B[12]=stif.x32/det;
  B[13]=stif.y23/det;
  B[14]=stif.x13/det;
  B[15]=stif.y31/det;
  B[16]=stif.x21/det;
  B[17]=stif.y12/det;
  }

void getD(STIF stif,double ym,double nu,double*D)
  {
  double c;
#if(defined(PlaneStress))
  c=ym/(1.0-nu*nu);
  D[0]=c;
  D[1]=nu*c;
  D[2]=0.;
  D[3]=nu*c;
  D[4]=c;
  D[5]=0.;
  D[6]=0.;
  D[7]=0.;
  D[8]=(1.-nu)*0.5*c;
#elif(defined(PlaneStrain))
  c=ym/((1.+nu)*(1.-2.*nu));
  D[0]=(1.-nu)*c;
  D[1]=nu*c;
  D[2]=0.;
  D[3]=nu*c;
  D[4]=(1.-nu)*c;
  D[5]=0.;
  D[6]=0.;
  D[7]=0.;
  D[8]=(0.5-nu)*c;
```

```
#else
  #error neither PlaneStrain nor PlaneStress
#endif
  }

void getKe(STIF stif,double ym,double nu,double
    tk,double*ke)
  {
  int i,j;
  double at,B[3*6],BT[6*3],BTD[6*3],D[3*3];
  getB(stif,B);
  Transpose(B,BT,3,6);
  getD(stif,ym,nu,D);
  Multiply(BT,D,BTD,6,3,3);
  Multiply(BTD,B,ke,6,3,6);
  at=tk*(stif.x13*stif.y23-stif.x23*stif.y13)/2.;
  for(i=0;i<6;i++)
    for(j=0;j<6;j++)
      ke[6*i+j]*=at;
  }
```

We build the global matrix by inserting the stiffness matrix for each element in the proper place for the nodes that form the element:

```
void Global(double*ke,int e)
  {
  int i,j,k,l,m[6];
  m[0]=(Elem[e].i)*DOF;
  m[1]=(Elem[e].i)*DOF+1;
  m[2]=(Elem[e].j)*DOF;
  m[3]=(Elem[e].j)*DOF+1;
  m[4]=(Elem[e].k)*DOF;
  m[5]=(Elem[e].k)*DOF+1;
  for(i=0;i<6;i++)
    {
    k=m[i];
    for(j=0;j<6;j++)
      {
      l=m[j]-k;
      if(l>=0)
        globe(k,l)+=ke[6*i+j];
      }
    }
  }
```

We add the point loads as forces on the respective nodes in the solution vector:

```
if(loads)
  {
  printf("  applying poing loads\n");
```

```
    for(i=0;i<loads;i++)
      {
      solu[(Load[i].n-1)*DOF]+=Load[i].Fx;
      solu[(Load[i].n-1)*DOF+1]+=Load[i].Fy;
      }
    }
```

We add the bearings (horizontal and/or vertical constraints) using the penalty method. This is a round-about way of saying that we connect a really stiff spring between the node and its original location.

```
  if(bears)
    {
    printf("  applying bearings using penalty
   method\n");
    pen=globe(1,0);
    for(i=1;i<nvar;i++)
      pen=max(pen,globe(i,0));
    pen=log10(pen);
    pen=pow(10.,ceil(pen)+10.);
    for(i=0;i<bears;i++)
      {
      if(Bear[i].Rx)
        globe((Bear[i].n-1)*DOF,0)*=pen;
      if(Bear[i].Ry)
        globe((Bear[i].n-1)*DOF+1,0)*=pen;
      }
    }
```

We then solve the system of linear equations using Gauss Elimination. We must always use row pivoting with this method, but we can't use column pivoting in this case, because the matrix is sparse and we don't ever deal with the full matrix.

```
void Gauss(double*x)
  {
  int i,ii,j,k,l,m,n;
  double aik,akj,piv,sum;
  for(ii=0;ii<nvar-1;ii++)
    {
    piv=globe(ii,0);
    m=ii+span;
    if(m>nvar)
      m=nvar;
    for(i=ii+1;i<m;i++)
      {
      aik=globe(ii,i-ii);
      for(l=0;l<span-1;l++)
        {
        if(i-ii+l<span)
          {
```

```
            akj=globe(ii,i-ii+1);
            globe(i,l)-=aik/piv*akj;
            }
          }
        x[i]=x[i]-aik/piv*x[ii];
        }
      }
    piv=globe(nvar-1,0);
    x[nvar-1]=x[nvar-1]/piv;
    for(k=nvar-2;k>=0;k--)
      {
      sum=0.;
      n=span-1;
      if(n+k>=nvar)
        n=nvar-k-1;
      for(j=1;j<=n;j++)
        sum=sum+globe(k,j)*x[j+k];
      x[k]=1./globe(k,0)*(x[k]-sum);
      }
    }
```

After solving for the unknown stresses, we calculate the strains by applying the stiffness matrix to each element:

```
    for(i=0;i<elems;i++)
      {
      stif=Stiff(Node[Elem[i].i].x,Node[Elem[i].i].y,
                 Node[Elem[i].j].x,Node[Elem[i].j].y,
                 Node[Elem[i].k].x,Node[Elem[i].k].y);
      getB(stif,B);
      k[0]=Elem[i].i*DOF;
      k[1]=Elem[i].i*DOF+1;
      k[2]=Elem[i].j*DOF;
      k[3]=Elem[i].j*DOF+1;
      k[4]=Elem[i].k*DOF;
      k[5]=Elem[i].k*DOF+1;
      for(j=0;j<6;j++)
        disp[j]=defo[k[j]];
      Multiply(B,disp,Elem[i].st,3,6,1);
      getD(stif,Elem[i].ym,Elem[i].nu,D);
      Multiply(D,Elem[i].st,Elem[i].ss,3,3,1);
      }
```

That's all there is to it. There's an optional zoom factor to exaggerate the nodal point displacements for visualization. It defaults to 5, but you can specify it as a second argument when you launch the program. The first argument is the input file name. The simplest problem we will consider is the one prescribed in the sample input file above (canti.fem). This is a cantilever beam, rigidly attached on the left with a downward force on the right end. The resulting nodal point displacements are shown in this next figure:

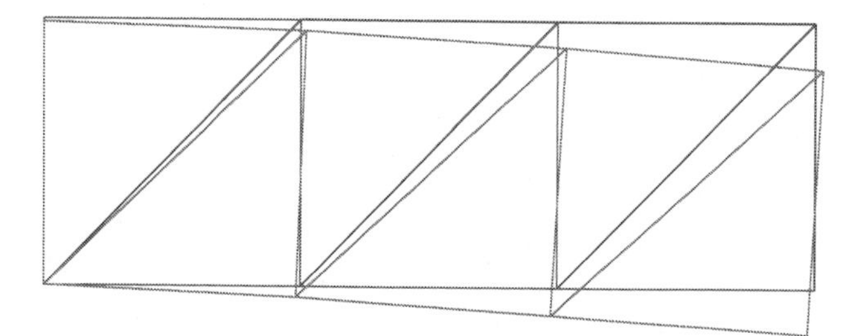

The blue lines are the undeformed elements and the red lines are the deformed ones, with an exaggeration factor (zoom) of 5. Output is written to two files. The first (psfem.out) contains the stresses and strains and the second (psfem.2dv) contains two *bodies* (a group of nodes and elements). The first is the undeformed and the second is the deformed with exaggeration. The next example is a simply-loaded beam (beam1.fem), supported at the ends with a downward force in the middle:

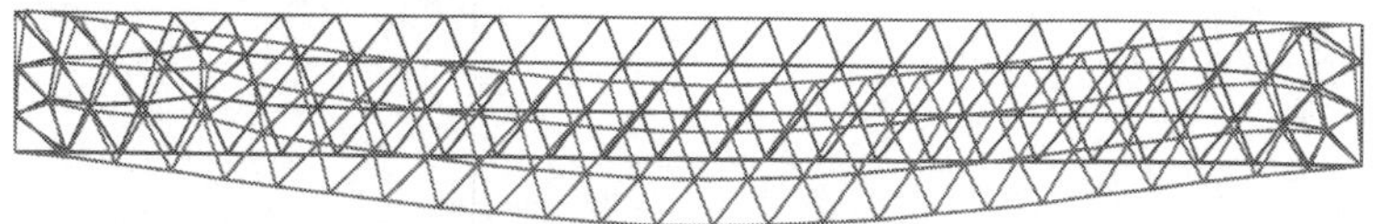

The next example is a much more finely discretized beam (beam2.fem) that is clamped at both ends with a load in the middle.

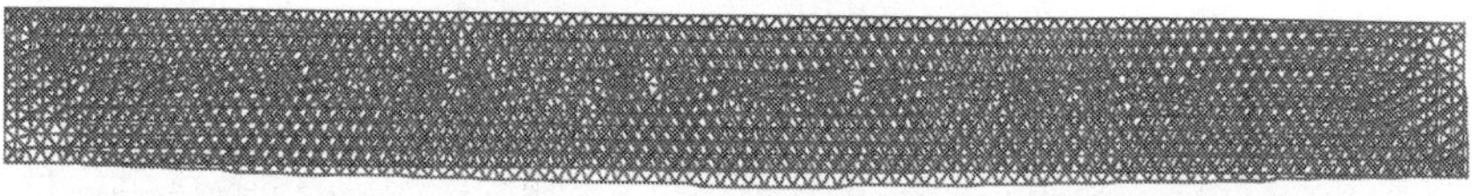

The next example is a flange with a hole in it that is stretched (flange.fem):

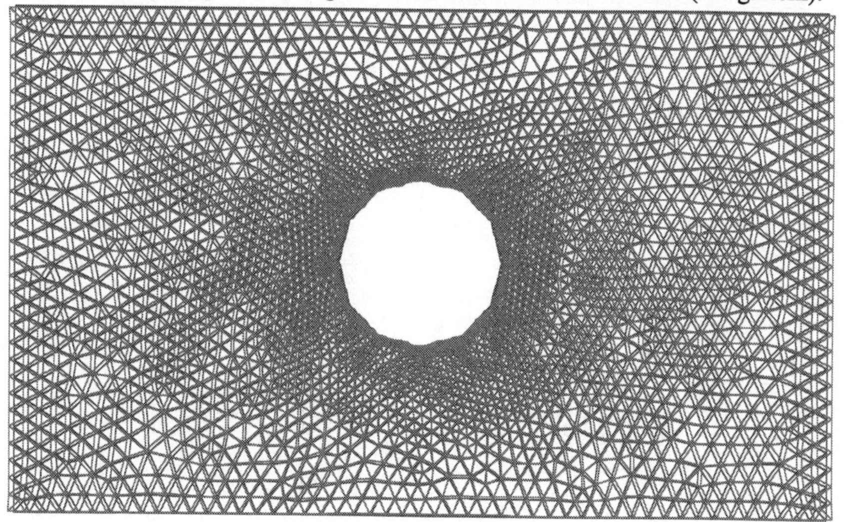

The last example is a loaded hook (hook.fem):

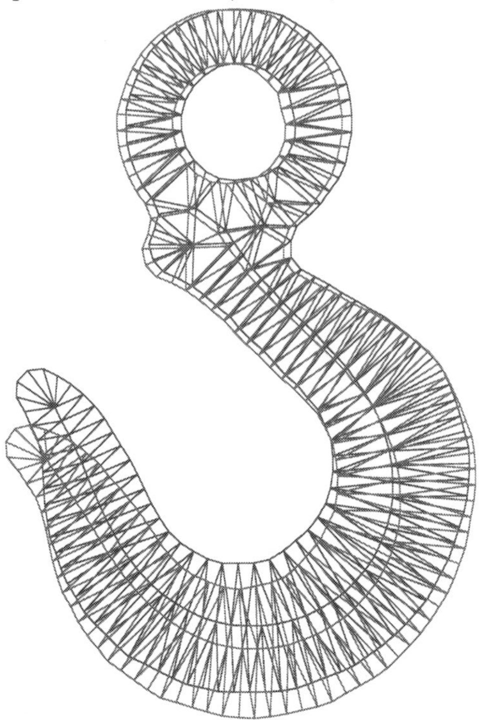

This is what the output of the program looks like. From this you can see the order of operations from reading the input file to summarizing the results:

```
Plane Strain Finite Element Model
reading input: hook.fem
  reading 238 nodes
  reading 298 elements
  reading 2 loads
  reading 2 bearings
begin solution
  sizing matrices
  applying point loads
  applying bearings using penalty method
  solving matrix for results
  calculating displacements
saving results: psfem.out
creating 2D view file: psfem.2dv
```

Chapter 7. Boundary Element Method

The boundary element method arises from Green's Lemma[14]. This remarkable relationship between the integral of a function over an area and the integral of a corresponding function around the perimeter of the same domain is usually covered in advanced calculus. Green's Lemma can be expressed by the following integral:

$$\iint \nabla^2 \varphi \, dA = \int \frac{\partial \varphi}{\partial n} dS \tag{7.1}$$

In Equation 7.1 φ is the potential, dA is the differential area within the domain, dS is a differential distance along the boundary, and n is the normal (perpendicular) at each location along the boundary. The potential could be simply that (i.e., electrostatic potential or invicid flow), temperature, concentration, stress, strain, or anything else that satisfies Laplace's equation.

We must also have a *fundamental* solution to Laplace's equation—in this case, a general, homogeneous solution (i.e., works for any case and is zero on the right-hand side). The fundamental solution is similar to the basis function postulated in the finite element method. Instead of satisfying the variational statement, this function must satisfy the equation itself. The derivation is a bit circuitous, except in polar coordinates.

$$\varphi = \ln\left(\frac{1}{r}\right) \tag{7.2}$$

We can at least show by substitution that Equation 7.2 satisfies Laplace's equation in polar coordinates:

$$\frac{1}{r}\frac{\partial}{\partial r}\left(r\frac{\partial \varphi}{\partial r}\right) + \frac{1}{r^2}\frac{\partial^2 \varphi}{\partial \theta^2} = 0 \tag{7.3}$$

For Cartesian (x,y) coordinates, r in the above equation is the distance from some as yet unspecified location $(x=a,\ y=b)$, or $r^2=(x-a)^2+(y-b)^2$. We further propose that the potential, φ, and the derivative with respect to the normal, $\partial\varphi/\partial n$, have some specific, finite, non-trivial value at each of the points along the boundary. Consider two points along the boundary (1 and 2). We can write Equation 7.2 for these two points:

[14] George Green (1729-1841): British mathematical physicist best known for work with electric fields and magnetism.

$$\varphi(r)=\left\{\frac{\varphi_1(\mathrm{r}-\mathrm{r}_2)}{\ln\left(\frac{1}{\mathrm{r}_1}\right)}-\frac{\varphi_2(\mathrm{r}-\mathrm{r}_1)}{\ln\left(\frac{1}{\mathrm{r}_2}\right)}\right\}\frac{\ln\left(\frac{1}{\mathrm{r}}\right)}{(\mathrm{r}_1-\mathrm{r}_2)} \tag{7.4}$$

At x_1,y_1 $r=r_1$ and at x_2,y_2 $r=r_2$ so that $\varphi(r_1)=\varphi_1$ and $\varphi(r_2)=\varphi_2$; therefore, Equation 7.4 is the particular form of the fundamental equation that satisfies Laplace's equation and matches at these two points along the boundary. We can construct a similar equation for the right side of Equation 7.1. When integrate Equation 7.4 from point 1 to point 2, we will get some constant times φ_1 plus some other constant times φ_2. We will use $H_{i,j}$ to represent these constants on the left side and $G_{i,j}$ to represent the corresponding constants on the right side of Equation 7.1. The indices *i* and *j* indicate each segment along the boundary and each pair corresponding to each segment. The resulting set of equations can be written:

$$\sum_{i=}^{n}\sum_{j=1}^{n}H_{i,j}\varphi_j=\sum_{i=}^{n}\sum_{j=1}^{n}G_{i,j}\frac{\partial\varphi_j}{\partial n} \tag{7.5}$$

Equation 7.5 constitutes a set of simultaneous linear equations for the potential function and its derivative at the *n* points along the boundary. We will need three different formulas for the integrals. We can readily integrate this equation around the boundary except at the two points (here, 1 and 2). At those points $r=r_1$ or $r=r_2$ and the standard result is indeterminate. We use a different formula for that one segment. We use these two formulas for points along the boundary. For points inside the boundary we use a third formula. We need this third integral to evaluate the results of the solution inside the boundary.

I have been vague up until this point on exactly what formulas are integrated and how, sparing you the gory details. In most applications, numerical integration (e.g., Gauss Quadrature) is used, because the analytical integral is unknown to the programmer. In fact, that's the way I present this as an example in my book *Numerical Calculus*. Here, you have the benefit of the analytical solution, which is precise, instead of a numerical solution that is approximate. Of course, I didn't figure this out the hard way. I used Maple® to do that for me. The resulting formulas are indeed tedious. The functions that accomplish these three integrals (adjacent, opposite, and internal) are:

```
void AdjacentBoundary(double X1,double Y1,double
   X2,double Y2,double*G1,double*G2)
  {
  double A,dS;
  dS=hypot(X2-X1,Y2-Y1);
  A=log(dS);
```

```
  *G1=dS*(1.5-A)/2;
  *G2=dS*(0.5-A)/2;
  }
void OppositeBoundary(double Xp,double Yp,double
   X1,double Y1,double X2,double
   Y2,double*G1,double*G2,double*H1,double*H2)
  {
  double pX=Xp-X1;
  double pY=Yp-Y1;
  double dX=X2-X1;
  double dY=Y2-Y1;
  double SS=dX*dX+dY*dY;
  double dS=sqrt(SS);
  double t3=-pX*dY+dX*pY;
  double t5=1/SS;
  double t6=t3*t5;
  double t7=pX*pX;
  double t8=pY*pY;
  double t10=log(t7+t8);
  double t12=-t3*t5;
  double t13=dX*dX;
  double t14=dX*pX;
  double t16=dY*dY;
  double t17=dY*pY;
  double t20=log(t13-2*t14+t7+t16-2*t17+t8);
  double t22=t13+t16-t17-t14;
  double t24=-1/t3;
  double t26=atan(-t22*t24);
  double t29=t17+t14;
  double t31=atan(t29*t24);
  double t42=1/SS/dS;
  double t43=t13*dX;
  double t45=2*t43*pX;
  double t46=t7*t13;
  double t47=t8*t16;
  double t48=t7*t16;
  double t49=t8*t13;
  double t51=4*t14*t17;
  double t52=t16*dY;
  double t54=2*t52*pY;
  double t57=2*t13*dY*pY;
  double t58=t16*dX;
  double t60=2*t58*pX;
  double t64=t13*t13;
  double t65=t13*t16;
  double t66=2*t65;
  double t67=t16*t16;
  double t68=t64+t46+t47-t48-t54+t66-t45+t67+t51-t57-
   t60-t49;
```

```
  double t74=dY*t8*dX;
  double t78=pX*pY*t13;
  double t80=t7*dX*dY;
  double t82=pX*t16*pY;
  double t84=-t52*pX+t43*pY-t74-t13*pX*dY-
    t78+t80+t82+t58*pY;
  double t101=t80+t82-t78-t74;
  *H1=-t6*t10/2-t12*t20/2-t22*t5*t26+t22*t5*t31;
  *H2=-t12*t10/2-t6*t20/2-t29*t5*t26+t29*t5*t31;
  *G1=-t42*(t45-t46-t47+t48+t49-t51+t54+t57+t60)*t10/4-
    t42*t68*t20/4-t42*t84*t26+t42*t84*t31-t42*(-6*t65-
    3*t64+t57+t60-3*t67+t45+t54)/4;
  *G2=t42*(t48-t46-t51-t47+t49)*t10/4+t42*(-t64-t67-
    t48+t47+t51+t46-t49-t66)*t20/4+t42*t101*t26-
    t42*t101*t31+t42*(t57+t60+t66+t45+t54+t64+t67)/4;
  }
int InternalIntegral(double Xp,double Yp,double
    X1,double Y1,double X2,double
    Y2,double*G1,double*G1x,double*G1y,double*G2,double*G
    2x,double*G2y,double*H1,double*H1x,double*H1y,double*
    H2,double*H2x,double*H2y)
  {
  double pX=Xp-X1;
  double pY=Yp-Y1;
  double dX=X2-X1;
  double dY=Y2-Y1;
  double SS=dX*dX+dY*dY;
  double dS=sqrt(SS);
  double t3=1/SS/dS;
  double t4=dY*dY;
  double t5=t4*t4;
  double t6=pY*pY;
  double t7=dX*dX;
  double t8=t6*t7;
  double t9=pX*pX;
  double t10=t9*t4;
  double t11=t4*dY;
  double t13=2*t11*pY;
  double t14=t7*dX;
  double t16=2*t14*pX;
  double t17=t7*t4;
  double t18=2*t17;
  double t19=t9*t7;
  double t20=t6*t4;
  double t21=t7*t7;
  double t22=pY*t7;
  double t24=2*t22*dY;
  double t25=t4*pX;
  double t27=2*t25*dX;
```

```
double t28=pX*dX;
double t29=pY*dY;
double t31=4*t28*t29;
double t32=t5-t8-t10-t13-t16+t18+t19+t20+t21-t24-
  t27+t31;
double t36=t7-2*t28+t9+t4-2*t29+t6;
double t37=log(t36);
double t41=t6+t9;
double t42=log(t41);
double t46=pY*dX;
double t48=dX*t9;
double t49=t48*dY;
double t50=pY*t4;
double t51=t50*pX;
double t52=t22*pX;
double t53=t7*pX;
double t55=dX*dY;
double t56=t55*t6;
double t58=4*t3*(t14*pY-pX*t11+t46*t4+t49+t51-t52-
  t53*dY-t56);
double t59=t7+t4-t28-t29;
double t61=t46-pX*dY;
double t62=1/t61;
double t64=atan(t59*t62);
double t66=t28+t29;
double t68=atan(t66*t62);
double t76=t55*pY;
double t77=2*t76;
double t78=t4*dX;
double t79=-t14+t53+t77-t78-t25;
double t84=2*t22;
double t85=t7*dY;
double t86=2*t85;
double t87=t28*dY;
double t88=4*t87;
double t89=2*t50;
double t90=2*t11;
double t92=t3*(-t84-t86+t88+t89-t90);
double t95=t14+t78;
double t98=2*t87;
double t99=-t22-t85+t98+t50-t11;
double t104=2*t14;
double t105=2*t53;
double t106=2*t78;
double t107=2*t25;
double t108=4*t76;
double t110=t3*(t104-t105+t106+t107-t108);
double t114=2*t3*(t85+t11);
double t123=4*t3*(t52+t56-t49-t51);
```

```
double t129=-t53-t77+t25;
double t135=t3*(t84-t88-t89);
double t140=-t22+t98+t50;
double t146=t3*(-t105-t108+t107);
double t150=1/SS;
double t151=-t61*t150;
double t153=t61*t150;
double t155=2*t59*t150;
double t159=dY*t6;
double t160=dY*t9;
double t161=t159+t160;
double t162=1/t41;
double t164=t150*t37;
double t167=t150*t42;
double t169=dX*t6;
double t170=-t169-t48;
double t171=2*t170*t162;
double t172=t150*t64;
double t174=t150*t68;
double t184=2*t161*t162;
double t193=2*t66*t150;
double t197=-t159-t160-t11-t85+t89+t98;
double t198=1/t36;
double t207=(t104+2*t169+2*t48+t106-4*t53-t108)*t198;
double t214=t14-t105+t169-t77+t78+t48;
double t225=(-t88-4*t50+t90+2*t160+t86+2*t159)*t150;
*G1=-t3*t32*t37/4-t3*(t13+t16+t10+t8-t19-t20-
 t31+t27+t24)*t42/4-t58*t64/4-t58*t68/4-t3*(t16+t13-
 3*t5+t24+t27-6*t17-3*t21)/4;
*G1x=-(t3*t79*t37/2-
 t3*t79*t42/2+t92*t64/2+t92*t68/2+t3*t95);
*G1y=-(t3*t99*t37/2-
 t3*t99*t42/2+t110*t64/2+t110*t68/2+t114/2);
*G2=-t3*(t18-t19+t8-t20+t10-t31+t5+t21)*t37/4-t3*(-
 t8+t19+t31+t20-t10)*t42/4-t123*t64/4-t123*t68/4-t3*(-
 t16-t13-t18-t27-t5-t21-t24)/4;
*G2x=-(t3*t129*t37/2-
 t3*t129*t42/2+t135*t64/2+t135*t68/2-t3*t95);
*G2y=-(-t3*t140*t37/2+t3*t140*t42/2-t146*t64/2-
 t146*t68/2-t114/2);
*H1=-t151*t37/2-t153*t42/2-t155*t64/2-t155*t68/2;
*H1x=-(t161*t162*t164/2-
 t161*t162*t167/2+t171*t172/2+t171*t174/2+(t22+t50)*t1
 62*t150);
*H1y=-(t170*t162*t164/2-t170*t162*t167/2-t184*t172/2-
 t184*t174/2-(t25+t53)*t162*t150);
*H2=-t153*t37/2-t151*t42/2-t193*t64/2-t193*t68/2;
```

```
*H2x=-(t197*t198*t164/2-
 t197*t198*t167/2+t207*t172/2+t207*t174/2+(t11+t85-
 t22-t50)*t198*t150);
*H2y=-(t214*t150*t198*t37/2-
 t214*t150*t198*t42/2+t225*t198*t64/2+t225*t198*t68/2+
 (t53+t25-t14-t78)*t150*t198);
return(0);
}
```

You read in the boundary points and the boundary conditions. You build the equations using the preceding integrals. Then you solve the simultaneous equations for *2n* unknowns (the potential and its derivative at each of the *n* points) and that's it! You use the third integral to find the value of the potential and its derivative at any point inside the boundary. You can then use some graphics program (e.g., TP2) to plot lines of constant potential throughout the domain, now that you know the value at however many points are desired. You can also calculate velocity vectors (or heat flux vectors or magnetic field vectors) from the potential and it's normal derivative that you got from the internal integral function.

You will find the program source code and 8 examples in the on-line archive in the folder examples\boundary element. The following is a typical input file (curl.bem):

```
Ye & McCorquodale's 270° bend
152 -5000 number of boundary nodes, number of internal
   calculation points
 0.00  0.00 0 0
 0.43  0.00 0 0
 0.86  0.00 0 0
 1.30  0.00 0 0
 1.73  0.00 0 0
 2.16  0.00 0 0
 2.59  0.00 0 0
 3.03  0.00 0 0
```

The output for this example is:

```
PFLOW/2.31: Boundary Element Method for Potential Fields
  by Dudley J. Benton, dudley.benton@gmail.com
input file: curl.bem
title: Ye & McCorquodale's 270° bend
number of boundary points .......... 152
number of internal points .......... -5000
allocating memory .................. 372 KB
reading boundary points ............ OK
total number of boundaries ......... 1
number of internal regions ......... 0
number of essential BCs ............ 152
number of normal BCs ............... 0
number of tangent BCs .............. 0
```

```
sorting boundaries ................. OK
essential boundary conditions ...... 152
number of simultaneous equations ... 152
allocating memory .................. 556 KB
nodal point equations .............. 152
resorting boundary conditions ...... OK
simultaneous equations ............. 152
reordering solution set ............ OK
calculating tangentials ............ OK
results output file ................ PFLOW.OUT
boundary output file ............... PFLOW.P2D
velocity output file ............... PFLOW.V2D
potential file ..................... PFLOW.TRI
plot file .......................... PFLOW.TP2
plot command file .................. PFLOW.TP2
for velocity vectors run TP2 PFLOW.TP2
for potential contours run TP2 PFLOW.TP2
```

The potential field inside the domain is:

The velocity vectors are:

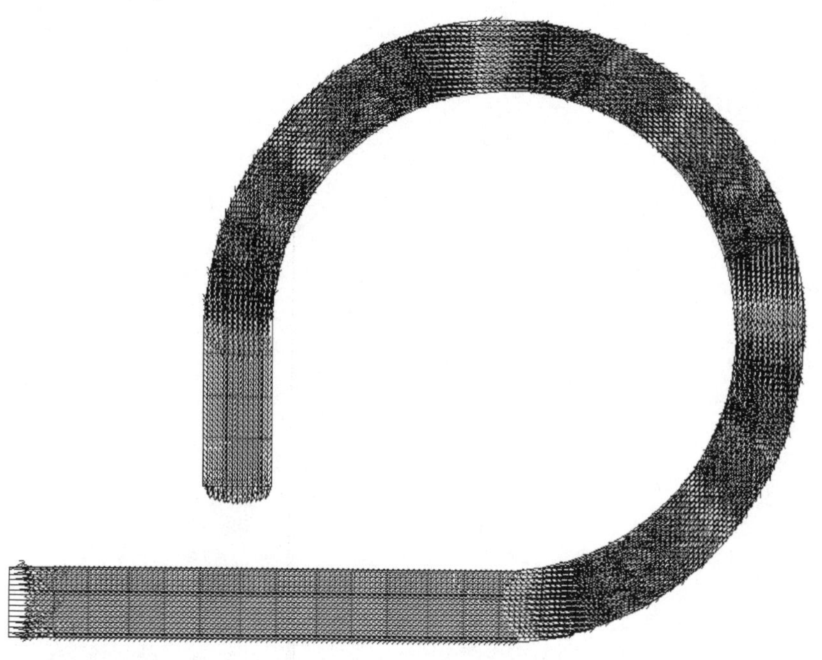

There's also a lake with two inlets, two outlets, and two islands (lake.bem). Potential is:

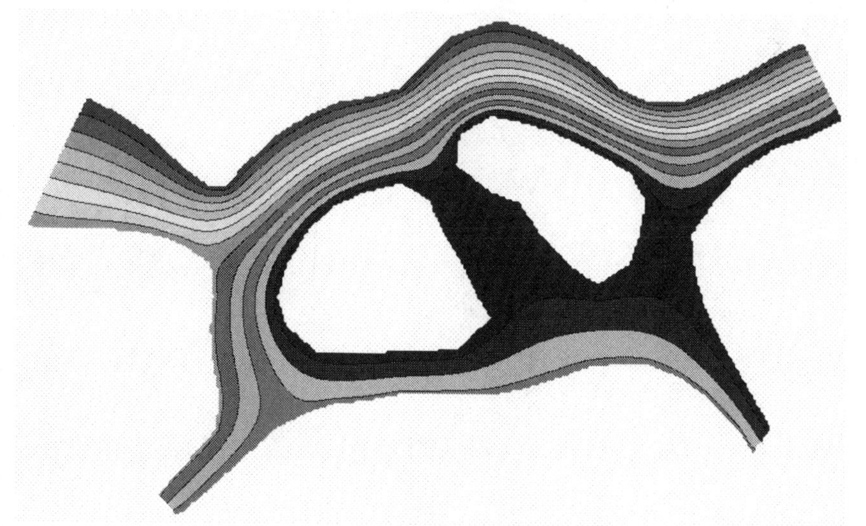

The velocity vectors are:

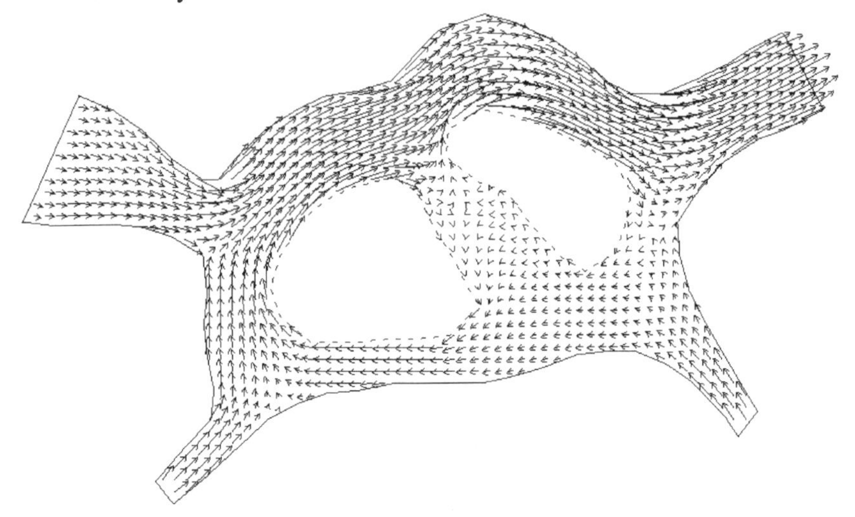

There is also a natural draft cooling tower in a slight cross wind (recirc.bem). The potential is:

The velocity vectors are:

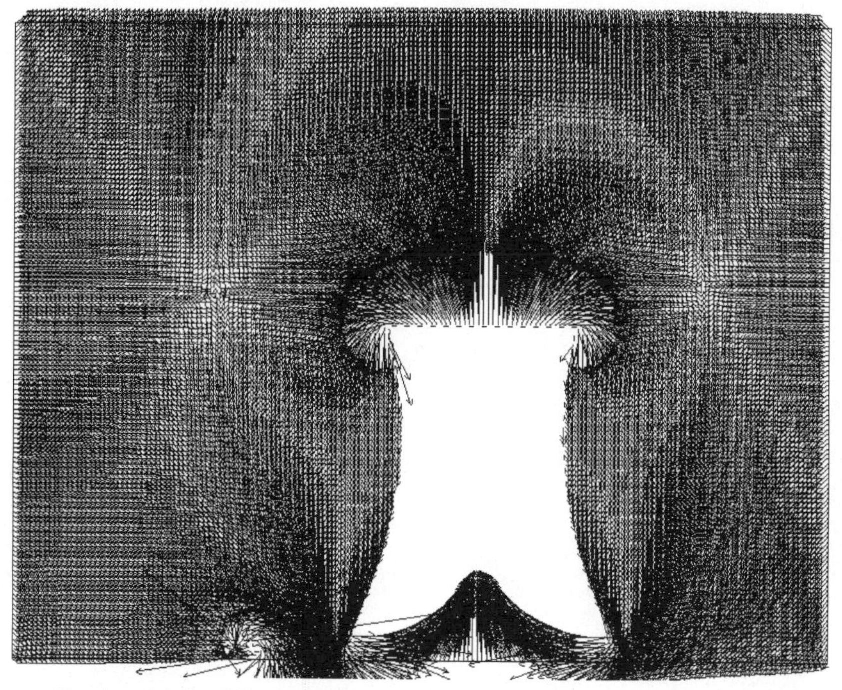

Here's a graph of flow through a coal-fired boiler (boiler.bem):

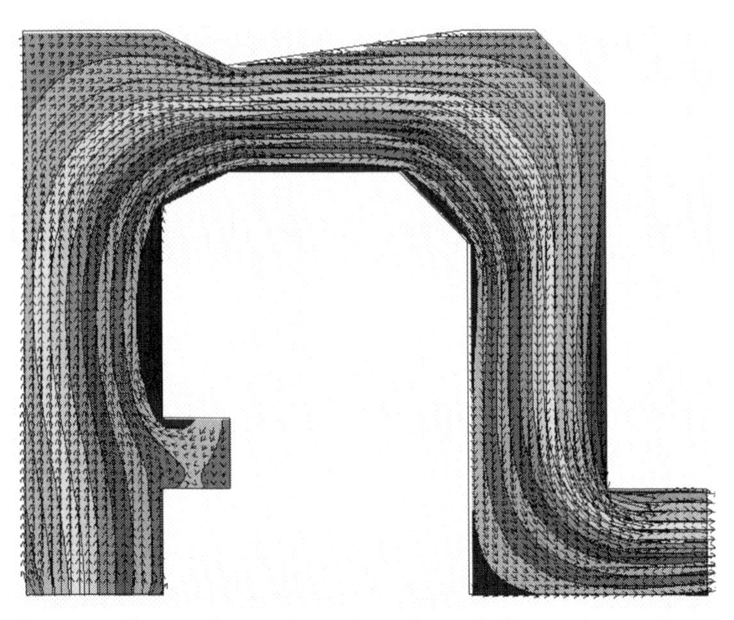

This next figure shows flow through a fork on the Holston River (reservior.bem):

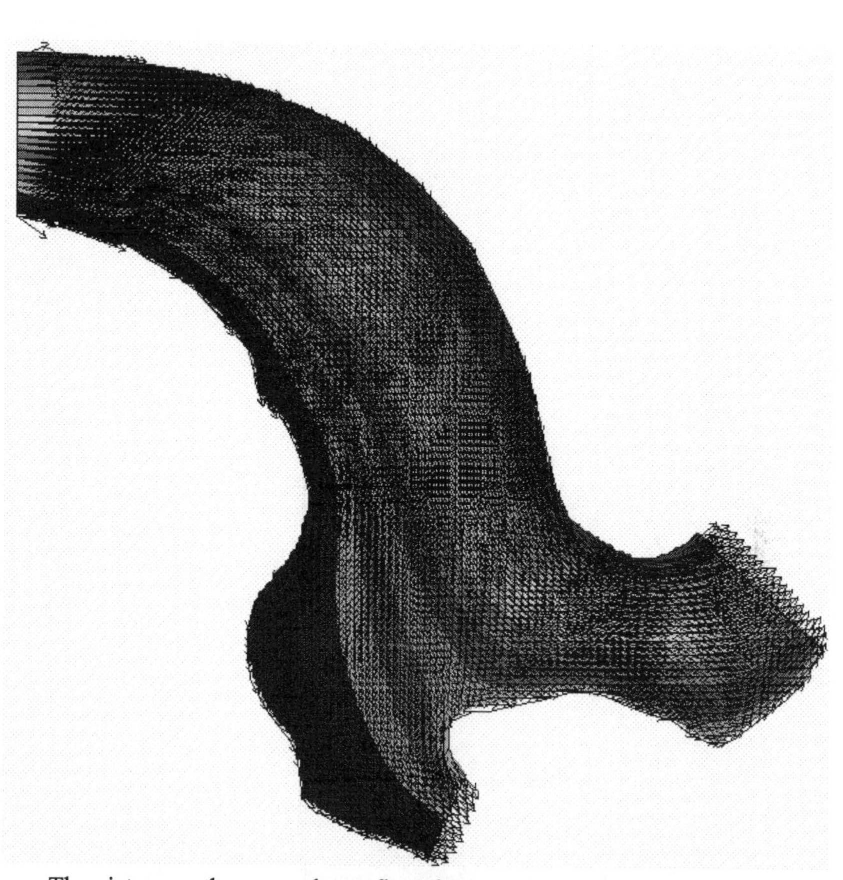

The picture on the cover shows flow through Tarpon Bend (bend.bem). You can use this method to calculate the flow through any shape.

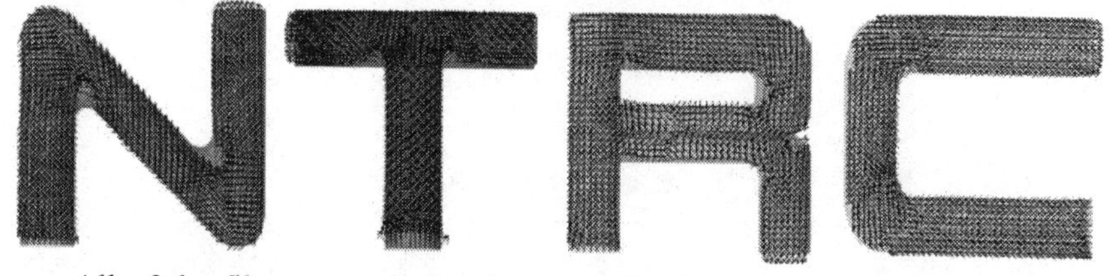

All of the files to create the above can be found in the on-line archive in folder examples\boundary element\NTRC.

Appendix A: Generalized Runge-Kutta Function

We will use the Butcher tableau form to define each of the methods. The following C code implements any of these:

```
/* Butcher tableau and working vector indexing macros */
#define a(i,j) BT[(steps+1)*i+1+j]
#define b(i)   BT[(steps+1)*steps+1+i]
#define c(i)   BT[(steps+1)*i]
#define k(i,j) W[n*(i+1)+j]
void RungeKutta(double*BT,int steps,void dYdX(double,
   double*,double*),double*X,double dX,double*Y,int n)
  {
  int i,s,step;
  static int m;
  static double*W;
  if(W==NULL||n*(steps+1)>m)
    {
    if(n*(steps+1)>m&&m>0)
      free(W);
    m=n*(steps+1);
    W=calloc(m,sizeof(double));
    }
  dYdX(*X,Y,&k(0,0));
  for(step=1;step<steps;step++)
    {
    for(i=0;i<n;i++)
      {
      W[i]=Y[i];
      for(s=0;s<step;s++)
        W[i]+=dX*a(step,s)*k(s,i);
      }
    dYdX(*X+dX*c(step),W,&k(step,0));
    }
  for(i=0;i<n;i++)
    for(s=0;s<steps;s++)
      Y[i]+=dX*b(s)*k(s,i);
  *X=*X+dX;
  }
```

This concise function allocates and preserves a working array in which to store the intermediate results. This array is expanded when necessary. When we are evaluating the performance of the different methods, we do not want to allocate and de-allocate an array every time the function is called, as this can take more time than the calculations and bias the results. This code can be found in the on-line archive in a file named RKcomparison.c. This is already compiled and will run on any version of the Windows® operating system, but there is also a batch file to recompile it, named: _compile_RKcomparison.bat. All of the tableaus can be found in this source code.

Appendix B: Finite Difference Operators

There are countless references to which you might look for guidance regarding finite difference operators. If you weren't already familiar with these concepts, you would not have gotten so far along with this book. There is one text that stands above the rest when it comes to reminding you of what you should have remembered, filling in the gaps, and taking you farther into the material and that is the *Handbook of Mathematical Functions* by Abramowitz and Stegun.[15] You can download a PDF of this 470 page treasure trove here:

http://people.math.sfu.ca/~cbm/aands/abramowitz_and_stegun.pdf

You will find everything you need to be reminded about finite difference operators in Chapter 25, "Numerical Interpolation, Differentiation, and Integration." I will only include one of the many such formulas available in this landmark reference, as I wouldn't want to deprive you of the experience of looking up the rest.

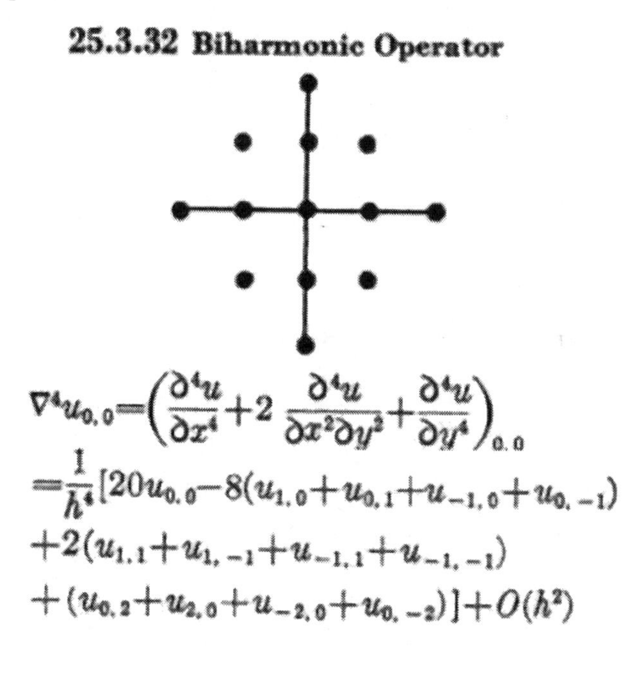

25.3.32 Biharmonic Operator

$$\nabla^4 u_{0,0} = \left(\frac{\partial^4 u}{\partial x^4} + 2\frac{\partial^4 u}{\partial x^2 \partial y^2} + \frac{\partial^4 u}{\partial y^4}\right)_{0,0}$$

$$= \frac{1}{h^4}[20u_{0,0} - 8(u_{1,0} + u_{0,1} + u_{-1,0} + u_{0,-1})$$

$$+2(u_{1,1} + u_{1,-1} + u_{-1,1} + u_{-1,-1})$$

$$+(u_{0,2} + u_{2,0} + u_{-2,0} + u_{0,-2})] + O(h^2)$$

[15] Abramowitz, M. and I. A. Stegun, *Handbook of Mathematical Functions* first published by the National Bureau of Standards as Technical Monograph No. 55. This very useful reference may be obtained free on-line as a PDF from several different web sites.

Appendix C: Processing and Checking Finite Element Models

There are several errors that can creep into a finite element model, including: unused nodes, coincident boundaries, overlapping boundaries, coincident nodes, elements having duplicate nodes, and degenerate elements. The program CHECK2D.c in folder examples\finite element\checker checks for all of these errors. Typical inputs are:

```
51 nodes
421.333 733.333
460.667 854.667
500 976
etc...
70 elements
2 3 4
8 9 10
26 27 28
etc...
```

Typical outputs are:

```
CHECK2D: check 2D elements
input file=star.2dv
reading nodes
  expecting 51 nodes
  allocating memory for nodes
  51 nodes found
  0≤X≤1000
  24≤Y≤976
reading elements
  expecting 70 elements
  allocating memory for elements
  70 elements found
checking for errors
  unused nodes... none
  coincident boundaries... none
  overlapping boundaries... none
  coincident nodes... none
  elements having duplicate nodes... none
  degenerate elements... none
elements are clockwise
```

This code contains several useful functions, including: polygon area and inside polygon test.

Appendix D: Solving Large Systems of Linear Equations

There are several methods available for solving large systems of linear equations. These are often sparse and have certain characteristics, including: being positive definite and occasionally symmetric. The two we will consider here are Successive Over Relaxation (SOR) and Conjugate Gradient (C-G). The SOR method is trivial to implement: simply initialize to something reasonable and then iterate until the changes become small. This method could be called: *Plug-and-Chug*. A relaxation factor of 1.5 is typical. The following code implements this calculation:

```
void SuccessiveOverRelaxation()
  {
  int i,iTx,j;
  double dT,dTx,Told,Tnew;
  dTx=0;
  iTx=0;
  converged=1;
  for(i=0;i<Nn;i++)
    {
    Told=Tn[i];
    Tnew=Bn[i];
    for(j=0;j<Na[i];j++)
      Tnew-=Tn[Ia[Ma*i+j]]*An[Ma*i+j];
    Tnew/=Dn[i];
    Tnew=(1.-Relax)*Told+Relax*Tnew;
    Tn[i]=Tnew;
    dT=fabs(Tnew-Told);
    if(dT>dTmax)
      converged=0;
    if(dT<=dTx)
      continue;
    iTx=i;
    dTx=dT;
    }
  }
```

The C-G method was developed by Hestenes and Stiefel.[16] It is also an iterative method. The approach is best presented in terms of vectors. The basic problem A[]X[]=B[] can be rearranged to produce a residual: B[]-A[]X[]=R[], where R[] is the residual obtained with each estimate of the solution, X[]. The sum of the squares of the residual is R^2=R^T[]R[], where R^T[] is the transpose of R[]. We define a step size along the direction between the previous and current estimates of the solution such that the R^2 is minimized. We then take another step that is orthogonal (perpendicular) the previous step and repeat the process.

[16] Hestenes, M. R. and Stiefel, E., "Methods Conjugate Gradients for Solving Linear Systems," *Journal of Research of the National Bureau of Standards*, Vol. 49, No. 6, Paper 2379, 1952.

If each step is orthogonal to all of the previous ones, there can be at most *n* steps, where *n* is the number of equations, because this is the total dimension of the space defined by the solution vectors. If the solution doesn't diverge due to round-off or ill condition of the matrix A[], the process must converge to the solution. This would be extremely burdensome and impractical if it weren't for the fact that an adequate solution is often achieved with only a few steps. The C-G method is truly remarkable in it's efficiency. Here's a simple function to implement the C-G method:

```
void ConjugateGradient(double*A,double*B,double*X,
    int*J,int n,int m)
  {
  int i,iter,k,l;
  double a,b,*D,dTq,q,*Q,r,*R,rTr;
  printf("begin Conjugate-Gradient\n");
  D=calloc(n,sizeof(double));
  Q=calloc(n,sizeof(double));
  R=calloc(n,sizeof(double));
  for(i=0;i<n;i++)
    X[i]=0.;
  for(rTr=l=i=0;i<n;i++)
    {
    r=B[i];
    for(k=0;k<m;k++,l++)
      if(J[l]>=0)
        r-=A[l]*X[J[l]];
    D[i]=R[i]=r;
    rTr+=r*r;
    }
  for(iter=0;;iter++)
    {
    for(dTq=l=i=0;i<n;i++)
      {
      for(q=k=0;k<m;k++,l++)
        if(J[l]>=0)
          q+=A[l]*D[J[l]];
      Q[i]=q;
      dTq+=D[i]*q;
      }
    a=rTr/dTq;
    for(i=0;i<n;i++)
      X[i]+=a*D[i];
    b=rTr;
    for(rTr=i=0;i<n;i++)
      {
      R[i]-=a*Q[i];
      rTr+=R[i]*R[i];
      }
    b=rTr/b;
```

```
    for(i=0;i<n;i++)
      D[i]=R[i]-b*D[i];
    if(iter%100==0)
      printf("iter=%li, rTr=%lG, a=%lG,
   b=%lG\n",iter,rTr,a,b);
    if(rTr<FLT_EPSILON)
      break;
    }
  free(D);
  free(Q);
  free(R);
  }
```

This code along with sample inputs can be found in the on-line archive in the folder examples\matrix in file conjugate_gradient.c. This code also contains a function for the Gauss-Siedel method, which is just the SOR method with the relaxation parameter set to 1. Typical output for both methods is:

```
begin Gauss-Siedel
iter=0, rTr=2.16423
iter=1, rTr=0.666334
iter=2, rTr=0.253819
iter=3, rTr=0.10727
iter=4, rTr=0.0472994
iter=5, rTr=0.0211947
...
iter=18, rTr=5.77393E-007
iter=19, rTr=2.55549E-007
iter=20, rTr=1.13083E-007
begin Conjugate-Gradient
iter=0, rTr=4.50617, a=0.555556, b=0.450617
iter=100, rTr=0.00476629, a=0.0421095, b=1.22691
iter=200, rTr=0.000302086, a=0.00838571, b=1.24215
iter=300, rTr=1.40871E-005, a=0.0125829, b=1.19524
iter=400, rTr=9.45145E-006, a=0.00536786, b=1.18726
iter=500, rTr=2.41304E-006, a=0.010531, b=1.14972
iter=600, rTr=3.11957E-007, a=0.00717648, b=1.15689
iter=700, rTr=1.58797E-007, a=0.00515067, b=1.15158
```

I chose marginally ill-conditioned A[] and initial estimate X[]={0} for the C-G method just to illustrate the point that this method will converge eventually, regardless of the initial estimate. Most examples converge much faster than this, including the following.

Vectorization

Some matrix solution methods can be vectorized and some can't. The Gauss-Siedel and SOR methods don't lend themselves to vectorization, or if they do, there is negligible improvement in speed. Gauss Elimination (GE) and the Conjugate-Gradient (C-G) methods to vectorize quite well. Some machines and operating systems come with vector processing instructions, while on others, this can be implemented with software, usually written in assembler. Some graphics cards provide vector processing and have special functions that you can call to access this power. When you're working with very large matrices or solving many problems sequentially, this can be of great advantage.

I have included a program with source code to implement vectorization for these two methods. You can find it in the folder examples\matrix. This program (vector.c) allocates arrays (you can change the size), fills them with a simple problem (Laplace's Equation with Dirichlet boundary conditions), and solves it using the various methods. Program output looks like:

```
Test Banded Matrix Solvers
by Dudley J. Benton
Solving Laplace's Equation
with Dirichlet boundary conditions
grid points: 8x8=64

solving using Gauss-Seidel
iter=1, resid=0.5
iter=2, resid=0.1875
iter=3, resid=0.11377
iter=4, resid=0.0830078
iter=5, resid=0.0622158
iter=30, resid=0.000301372
iter=31, resid=0.000244635
error reduction 0.00048927

solving using SOR
iter=1 resid=0.75
iter=2 resid=0.166992
iter=3 resid=0.0967551
iter=4 resid=0.0674985
iter=5 resid=0.0345428
iter=13 resid=0.000773403
iter=14 resid=0.000147819
error reduction 0.000197091

solving using vectorized Hestenes-Stiefel
(Conjugate-Gradient)
iter=1 resid=2
iter=2 resid=0.875
iter=3 resid=0.455696
iter=4 resid=0.471959
iter=5 resid=0.307282
iter=6 resid=0.257967
iter=7 resid=0.136661
iter=8 resid=0.0796246
iter=9 resid=0.0366493
```

```
iter=10 resid=0.0129086
iter=11 resid=0.00130093
error reduction 0.00020064

solution
node    G-S       SOR      HSV
1       1         1        1
11      0.908066           0.90853 0.908622
21      0.631672           0.632514         0.632472
22      0.499389           0.5      0.499978
23      0.300216           0.300517         0.300591
64      0         0        0

Testing Gauss Elimination
1 1  1    1    1    1   1
1 2  4    8   16   32   2
1 3  9   27   81  243   6
1 4 16   64  256 1024  24
1 5 25  125  625 3125 120
1 6 36  216 1296 7776 720
vectorized
  scalar  vector
1 -264     -264
2 611.8    611.8
3 -511.583 -511.583
4 198.625  198.625
5 -36.4167 -36.4167
6  2.575   2.575
```

The source code contains vector instructions implemented in C, including: add, subtract, multiply, divide, dot product, and swap.

```
void vadd(double*v1,int incr1,double*v2,int
   incr2,double*v3,int incr3,int n)
  { /* v3=v1+v2 */
  int i1,i2,i3,j;
  for(i1=i2=i3=j=0;j<n;j++)
    {
    v3[i3]=v1[i1]+v2[i2];
    i1+=incr1;
    i2+=incr2;
    i3+=incr3;
    }
  }
void vsub(double*v1,int incr1,double*v2,int
   incr2,double*v3,int incr3,int n)
  { /* v3=v1-v2 */
  int i1,i2,i3,j;
  for(i1=i2=i3=j=0;j<n;j++)
    {
```

```
      v3[i3]=v1[i1]-v2[i2];
      i1+=incr1;
      i2+=incr2;
      i3+=incr3;
      }
    }
  void vmpy(double*v1,int incr1,double*v2,int
     incr2,double*v3,int incr3,int n)
    { /* v3=v1*v2 */
    int i1,i2,i3,j;
    for(i1=i2=i3=j=0;j<n;j++)
      {
      v3[i3]=v1[i1]*v2[i2];
      i1+=incr1;
      i2+=incr2;
      i3+=incr3;
      }
    }
  double vdot(double*v1,int incr1,double*v2,int incr2,int
     n)
    { /* v1 dot v2 */
    int i1,i2,j;
    double d;
    for(d=i1=i2=j=0;j<n;j++)
      {
      d+=v1[i1]*v2[i2];
      i1+=incr1;
      i2+=incr2;
      }
    return(d);
    }
  void vswp(double*v1,int incr1,double*v2,int incr2,int n)
    { /* vector swap */
    int i,i1,i2;
    double v;
    for(i1=i2=i=0;i<n;i++)
      {
      v=v1[i1];
      v1[i1]=v2[i2];
      v2[i2]=v;
      i1+=incr1;
      i2+=incr2;
      }
    }
```

The scalar C-G method has already been discussed and can be found in conjugate_gradient.c. The vectorized version for a span of 5 is listed below.

```
int Hest5(double*a1,double*a2,double*a3,double*a4,
    double*a5,double*b,double*x,double*r,double*p,
    double*ap,double*w,int nx,int nn,double*r1,double*r2)
  {
  int i,iter,nnx;
  double alpha,epsilon,pap,pr,rap;
  /* vectorized Hestenes-Stiefel (Conjugate Gradient)
     method for linear simultaneous equations */
  if(nn<=nx||nx<3||nn<9)
    return(OTHER);
  nnx=nn-nx;
  alpha=0.;
  vset(0.,p,1,nn);
  vset(0.,r,1,nn);
  i=1+vmib(a3,1,nn);
  if(fabs(a3[i-1])<DBL_EPSILON)
    return(SINGULAR);
  vdiv(b,1,a3,1,x,1,nn);
  for(iter=0;iter<nn;iter++)
    {
    if(iter==0)
      {
      vset(0.,r,1,nn);
      vsub(r,1,b,1,r,1,nn);
      vmpy(&a1[nx],1,x,1,&w[nx],1,nnx);
      vadd(&r[nx],1,&w[nx],1,&r[nx],1,nnx);
      vmpy(a2,1,&x[nx],1,w,1,nnx);
      vadd(r,1,w,1,r,1,nnx);
      vmpy(a3,1,x,1,w,1,nn);
      vadd(r,1,w,1,r,1,nn);
      vmpy(a4,1,&x[1],1,w,1,nn-1);
      vadd(r,1,w,1,r,1,nn-1);
      vmpy(&a5[1],1,x,1,&w[1],1,nn-1);
      vadd(&r[1],1,&w[1],1,&r[1],1,nn-1);
      }
    else
      vpiv(epsilon,ap,1,r,1,r,1,nn);
    i=1+vmab(r,1,nn);
    *r2=fabs(r[i-1]);
    if(iter==0)
      *r1=*r2;
    if(*r2<0.0005*(*r1))
      return(NOERROR);
    if(iter!=0)
      {
      rap=vdot(r,1,ap,1,nn);
      alpha=rap/pap;
      }
    vsmy(alpha,p,1,p,1,nn);
```

```
    vsub(p,1,r,1,p,1,nn);
    vset(0.,ap,1,nn);
    vmpy(&a1[nx],1,p,1,&w[nx],1,nnx);
    vadd(&ap[nx],1,&w[nx],1,&ap[nx],1,nnx);
    vmpy(a2,1,&p[nx],1,w,1,nnx);
    vadd(ap,1,w,1,ap,1,nnx);
    vmpy(a3,1,p,1,w,1,nn);
    vadd(ap,1,w,1,ap,1,nn);
    vmpy(a4,1,&p[1],1,w,1,nn-1);
    vadd(ap,1,w,1,ap,1,nn-1);
    vmpy(&a5[1],1,p,1,&w[1],1,nn-1);
    vadd(&ap[1],1,&w[1],1,&ap[1],1,nn-1);
    pap=vdot(p,1,ap,1,nn);
    if(fabs(pap)<DBL_EPSILON)
      return(SINGULAR);
    pr=vdot(p,1,r,1,nn);
    epsilon=-pr/pap;
    printf("iter=%i resid=%lG\n",iter+1,*r2);
    vpiv(epsilon,p,1,x,1,x,1,nn);
    }
  return(CONVERGENCE);
  }
```

The vectorized GE method is listed below. Both vectorized and scalar methods can be found in vector.c. The vectorizd version is implemented with conditional compilation directives (#ifdef... #else... #endif) to better illustrate the differences.

```
int GaussVector(double*A,double*B,int n)
  { /* vectorized Gauss elimination with row and column
    pivoting */
  int i,j,k,ip,jp,*pivot;
  double a,b,p;
  if(n<1)
    return(OTHER);
  if(n==1)
    {
    a=A[0];
    if(fabs(a)<FLT_MIN)
      return(SINGULAR);
    B[0]/=a;
    return(NOERROR);
    }
  if((pivot=calloc(n,sizeof(int)))==NULL)
    return(OTHER);
  for(i=0;i<n;i++)
    pivot[i]=i;
  for(k=0;k<n-1;k++)
    {
#ifndef vectorized
```

```
    ip=k;
    jp=k;
    p=fabs(A[n*k+k]);
    for(i=k;i<n;i++)
      {
      for(j=k;j<n;j++)
        {
        a=fabs(A[n*i+j]);
        if(a>p)
          {
          ip=i;
          jp=j;
          p=a;
          }
        }
      }
#else
    jp=n*k+k+vmab(A+n*k+k,1,n*(n-k)-k);
    ip=jp/n;
    jp-=n*ip;
    p=fabs(A[n*ip+jp]);
#endif
    if(p<FLT_MIN)
      {
      free(pivot);
      return(SINGULAR);
      }
    if(ip!=k) /* row pivot */
      {
#ifndef vectorized
      b=B[k];
      B[k]=B[ip];
      B[ip]=b;
#else
      vswp(B+k,1,B+ip,1,1);
#endif
#ifndef vectorized
      for(j=k;j<n;j++)
        {
        a=A[n*k+j];
        A[n*k+j]=A[n*ip+j];
        A[n*ip+j]=a;
        }
#else
      vswp(A+n*k+k,1,A+n*ip+k,1,n-k);
#endif
      }
    if(jp!=k) /* column pivot */
      {
```

```
      j=pivot[jp];
      pivot[jp]=pivot[k];
      pivot[k]=j;
#ifndef vectorized
      for(i=0;i<n;i++)
        {
        a=A[n*i+k];
        A[n*i+k]=A[n*i+jp];
        A[n*i+jp]=a;
        }
#else
      vswp(A+k,n,A+jp,n,n);
#endif
      }
#ifndef vectorized
    for(i=k+1;i<n;i++) /* row elimination */
      {
      a=-A[n*i+k]/A[n*k+k];
      B[i]+=a*B[k];
      for(j=k+1;j<n;j++)
        A[n*i+j]+=a*A[n*k+j];
      }
#else
    for(i=k+1;i<n;i++) /* row elimination */
      {
      a=-A[n*i+k]/A[n*k+k];
      B[i]+=a*B[k];
      vpiv(a,A+n*k+k+1,1,A+n*i+k+1,1,A+n*i+k+1,1,n-k-1);
      A[n*i+k]=0.; /* must put 0s in lower triangle for
   vmab() to work */
      }
#endif
    }
  a=A[n*n-1];
  p=fabs(a);
  if(p<FLT_MIN)
    {
    free(pivot);
    return(SINGULAR);
    }
  B[n-1]/=a;
  for(k=1;k<n;k++) /* backsolve */
    {
    i=n-1-k;
#ifndef vectorized
    b=0.;
    for(j=i+1;j<n;j++)
      b+=A[n*i+j]*B[j];
#else
```

```
      b=vdot(A+n*i+i+1,1,B+i+1,1,n-i-1);
#endif
      B[i]=(B[i]-b)/A[n*i+i];
      }
  for(i=0;i<n;i++) /* column unpivot */
    A[i]=B[i];
  for(i=0;i<n;i++)
    B[pivot[i]]=A[i];
  free(pivot);
  return(NOERROR);
  }
```

Besides replacing the for() loops with vector function calls, the only other change is that you must zero-out the lower part of the diagonal matrix in order for the vectorized pivot search to work properly. There's a comment on that line in the code. The vectorized pivot searches the entire bottom of the matrix in a single sweep. The scalar pivot only searches the lower right part of the matrix. If you don't put zeroes in lower part after you eliminate each row, the vector search will include these values in the search. It's not necessary to put zeroes in the lower part when performing the scalar search.

Appendix E: Triangular Element Generating Program

I have also included a small program (elem3.c) that will generate triangular elements from a closed polygon. It's very simple to use and all the files—including the source code—can be found in the on-line archive in the folder examples\grid generator. A typical polygon file looks like this:

```
POLY P1
0.510942647 -0.0510424392
0.549024467 -0.0889642514
0.600066906 -0.100004779
0.650949338 -0.0889642514
0.800076464 0
0.700071685 0
etc...
END
```

Which plots up with TP2 like this:

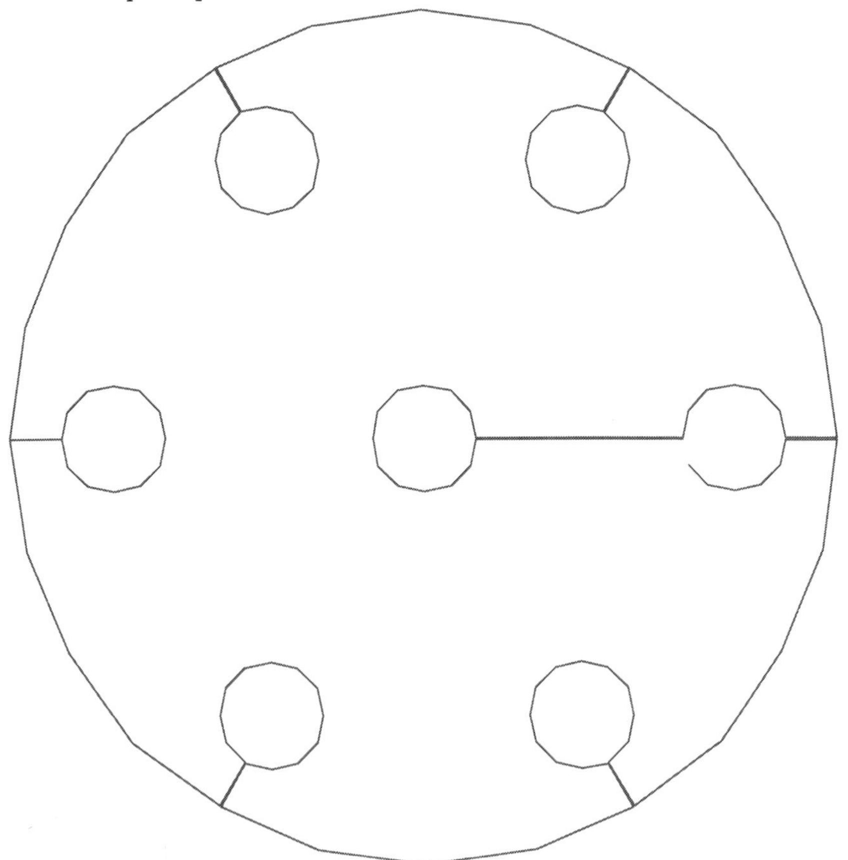

You have to split the lines that go into the bolt holes and then come back out to the edge. After processing it looks like this:

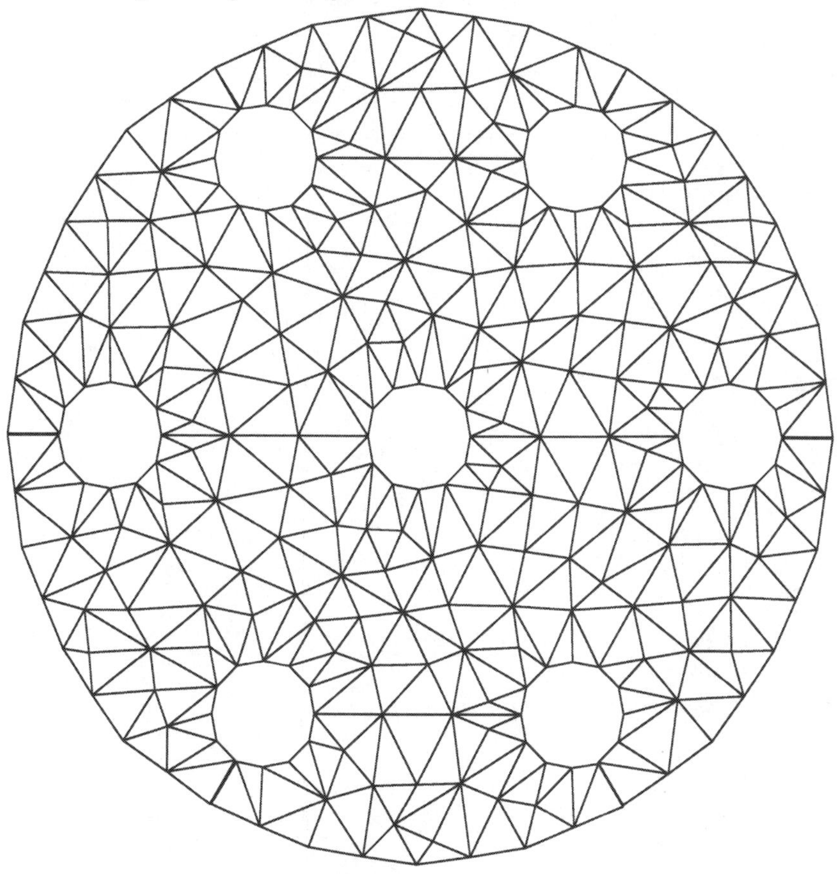

Typical output from the program is:

```
>elem3 lid.p2d
ELEM3/V2.00: Triangularization of a Closed Polygon
input file: lid.p2d
  scanning input file
  124 lines read
  122 boundary points found
  allocating memory
  reading input file
analyzing boundary polygon
  initializing points to boundary
  shortening sides
120 nodes, 124 elements
output file: lid.2dv
```

You can also control the size of the elements by specifying a target length as a second parameter after the input file name.

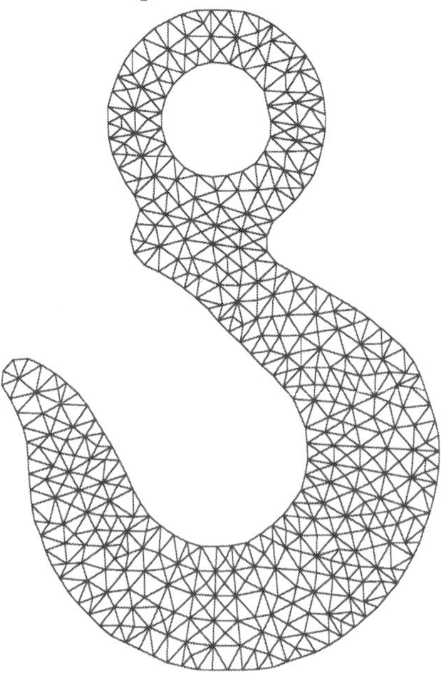

also by D. James Benton

3D Rendering in Windows: How to display three-dimensional objects in Windows with and without OpenGL, ISBN-9781520339610, Amazon, 2016.

Curve-Fitting: The Science and Art of Approximation, ISBN-9781520339542, Amazon, 2016.

Evaporative Cooling: The Science of Beating the Heat, ISBN-9781520913346, Amazon, 2017.

Heat Exchangers: Performance Prediction & Evaluation, ISBN-9781973589327, Amazon, 2017.

Jamie2 2nd Ed.: Innocence is easily lost and cannot be restored, ISBN-9781520339375, Amazon, 2016-18.

Little Star 2nd Ed.: God doesn't do things the way we expect Him to. He's better than that! ISBN-9781520338903, Amazon, 2015-17.

Living Math: Seeing mathematics in every day life (and appreciating it more too), ISBN-9781520336992, Amazon, 2016.

Lost Cause: If only history could be changed…, ISBN-9781521173770. Amazon 2017.

Mill Town Destiny: The Hand of Providence brought them together to rescue the mill, the town, and each other, ISBN-9781520864679, Amazon, 2017.

Monte Carlo Simulation: The Art of Random Process Characterization, ISBN-9781980577874, Amazon, 2018.

Numerical Calculus: Differentiation and Integration, ISBN-9781980680901, Amazon, 2018.

ROFL: Rolling on the Floor Laughing, ISBN-9781973300007, Amazon, 2017.

A Synergy of Short Stories: The whole may be greater than the sum of the parts, ISBN-9781520340319, Amazon, 2016.

Thermodynamics - Theory & Practice: The science of energy and power, ISBN-9781520339795, Amazon, 2016.

Version-Independent Programming: Code Development Guidelines for the Windows® Operating System, ISBN-9781520339146, Amazon, 2016.

Made in the USA
Lexington, KY
16 January 2019